AF360748

Juvenile Onset Huntington's Disease

Juvenile Onset Huntington's Disease

Editor

Peggy C. Nopoulos

MDPI • Basel • Beijing • Wuhan • Barcelona • Belgrade • Manchester • Tokyo • Cluj • Tianjin

Editor
Peggy C. Nopoulos
University of Iowa
USA

Editorial Office
MDPI
St. Alban-Anlage 66
4052 Basel, Switzerland

This is a reprint of articles from the Special Issue published online in the open access journal *Brain Sciences* (ISSN 2076-3425) (available at: https://www.mdpi.com/journal/brainsci/special_issues/Juvenile_Huntington_Disease).

For citation purposes, cite each article independently as indicated on the article page online and as indicated below:

LastName, A.A.; LastName, B.B.; LastName, C.C. Article Title. *Journal Name* **Year**, *Volume Number*, Page Range.

ISBN 978-3-03943-811-2 (Hbk)
ISBN 978-3-03943-812-9 (PDF)

Contents

About the Editor

Peggy C. Nopoulos is a professor of Psychiatry, Neurology, and Pediatrics. As of 2018, Dr. Nopoulos became the first female Chair and DEO of the Department of Psychiatry at the University of Iowa Hospital & Clinics. She received her M.D. degree from the University of Iowa in 1989 and completed her postdoctoral fellowship in Neuropsychiatry. Dr. Nopoulos' clinical and research experience in Huntington's Disease (HD) is broad and deep. She has been providing clinical care to patients with HD as part of the HD Center of Excellence since 2003. Early in her research career, Dr. Nopoulos worked as a co-investigator in a large neuroimaging study of individuals in the pre-symptomatic phases of HD that included 32 sites around the world. In 2009, Dr. Nopoulos embarked on a unique research endeavor called Kids-HD, with the aim to address brain development in children who inherited the adult-onset form of HD. Dr. Nopoulos expanded this project in 2011 to include children with juvenile-onset HD (JOHD). The Kids-JHD study constitutes the only prospective study in the world and this work has provided novel insights into the neurobiology of JOHD.

Preface to "Juvenile Onset Huntington's Disease"

Receiving a diagnosis of Huntington's Disease (HD) can be devastating for patients and families. However, the HD community has demonstrated their strength and resolve by engaging in clinical research efforts that have allowed researchers to better understand the course of HD, identify biomarkers, and investigate how various compounds impact the progression of the disease. Thanks to these efforts, the HD community has something that they have been desperately seeking for years: hope. Hope is being provided in the form of the first clinical trials aimed at slowing the progression of the disease. While these efforts are still in their infancy, simply knowing that advances are being made has provided hope to a desperate group of patients. Unfortunately, patients with Juvenile-Onset HD (JOHD) and their families may still be having trouble getting too excited about these potential therapeutic advances. There is still so much that is unknown about patients with JOHD that the HD community has been unable to make scientific advances with the same speed in this small group of patients. This book is meant to help bridge the gaps in knowledge that still remain about JOHD in an effort to provide therapeutic advances to those patients suffering from this disease. This book contains seven articles from authors around the globe who are engaging in clinical research to better understand the biology of JOHD. The information contained in this special edition represents some of the most cutting-edge information regarding JOHD. Our hope is that sharing this information will continue to broaden the HD community's knowledge about JOHD in order to advance therapeutic development and provide the same hope to these patients that patients with HD have been given in recent years.

Peggy C. Nopoulos
Editor

Editorial

Special Issue: Juvenile Onset Huntington's Disease

Peg C. Nopoulos [1,2,3]

1   Department of Psychiatry, Carver College of Medicine at the University of Iowa, Iowa City, IA 52242 USA;
     Peggy-Nopoulos@uiowa.edu; Tel.: +1-319-356-1144
2   Department of Neurology, Carver College of Medicine at the University of Iowa, Iowa City, IA 52242, USA
3   Department of Pediatrics, Carver College of Medicine at the University of Iowa, Iowa City, IA 52242, USA

Received: 14 September 2020; Accepted: 14 September 2020; Published: 20 September 2020

The Special Issue "Juvenile Onset Huntington's Disease" highlights the growing interest in understanding the unique aspects of this ultra-rare disorder. Adult Onset Huntington's Disease (AOHD) is a single gene disorder caused by a triplet repeat expansion in the Huntingtin gene. With decades of research to support the search for a cure, we are now in an exciting time of true progress in fighting AOHD with gene therapy trials underway. However, excluded from current studies are the subset of patients who, by virtue of very high CAG repeat expansion (typically over 60), have onset of disease early in life, defined by motor onset prior to age 21 and referred to as Juvenile Onset Huntington's Disease (JOHD). This definition is somewhat arbitrary as the pathogenic mechanism is exactly the same—expanded CAG repeat in the Huntingtin gene. Nevertheless, due to its rarity, there is a relative dearth of studies on JOHD, leaving many questions regarding its phenomenology.

The current issue includes seven articles that span a variety of topics including the difficult emotional experience that parents endure in the context of their child becoming ill and diagnosed with JOHD [1]; a review of the clinical manifestations of JOHD [2]; and four articles from the only prospective study of JOHD evaluating behavior [3], the association of CAG repeat and motor onset [4], autonomic nervous system dysfunction [5], and abnormality in the unique MRI marker of T1rho in JOHD subjects [6]. Finally, this issue is rounded out by a review of the therapeutic advances for HD, highlighting the possibilities in the future of the types of clinical trials that JOHD subjects may be included in [7].

The entire HD community—patients, family members at-risk for HD, caregivers, health-care professionals and scientists—has a keen interest in focusing attention on JOHD. There is a calling to better understand, and help, the plight of those that seem to have been "left behind" in the flurry of research studies on AOHD [8]. The study of patients who are afflicted early in life with HD has become an urgent need with this Special Issue representing just the beginning of the required effort.

References

1.   Oosterloo, M.; Bijlsma, E.K.; Die-Smulders, C.; Roos, R.A.C. Diagnosing Juvenile Huntington's Disease: An Explorative Study among Caregivers of Affected Children. *Brain Sci.* **2020**, *10*, 155. [CrossRef] [PubMed]
2.   Achenbach, J.; Thiels, C.; Lucke, T.; Saft, C. Clinical Manifestation of Juvenile and Pediatric HD Patients: A Retrospective Case Series. *Brain Sci.* **2020**, *10*, 340. [CrossRef] [PubMed]
3.   Langbehn, K.E.; Cochran, A.M.; van der Plas, E.; Conrad, A.L.; Epping, E.; Martin, E.; Espe-Pfeifer, P.; Nopoulos, P. Behavioral Deficits in Juvenile Onset Huntington's Disease. *Brain Sci.* **2020**, *10*, 543. [CrossRef] [PubMed]
4.   Schultz, J.L.; Moser, A.D.; Nopoulos, P.C. The Association between CAG Repeat Length and Age of Onset of Juvenile-Onset Huntington's Disease. *Brain Sci.* **2020**, *10*, 575. [CrossRef] [PubMed]
5.   Schultz, J.L.; Nopoulos, P.C. Autonomic Changes in Juvenile-Onset Huntington's Disease. *Brain Sci.* **2020**, *10*, 589. [CrossRef] [PubMed]

6. Tereshchenko, A.V.; Schultz, J.L.; Kunnath, A.J.; Bruss, J.E.; Epping, E.A.; Magnotta, V.A.; Nopoulos, P.C. Subcortical T1-Rho MRI Abnormalities in Juvenile-Onset Huntington's Disease. *Brain Sci.* **2020**, *10*, 533. [CrossRef] [PubMed]

7. Kumar, A.; Kumar, V.; Singh, K.; Kumar, S.; Kim, Y.S.; Lee, Y.M.; Kim, J.J. Therapeutic Advances for Huntington's Disease. *Brain Sci.* **2020**, *10*, 43. [CrossRef] [PubMed]

8. Stout, J.C. Juvenile Huntington's disease: Left behind? *Lancet Neurol.* **2018**, *17*, 932–933. [CrossRef]

Communication

Autonomic Changes in Juvenile-Onset Huntington's Disease

Jordan L. Schultz [1,2,*] and Peg C. Nopoulos [1,2,3]

[1] Department of Psychiatry, Carver College of Medicine at the University of Iowa, Iowa City, IA 52242, USA; peggy-nopoulos@uiowa.edu
[2] Department of Neurology, Carver College of Medicine at the University of Iowa, Iowa City, IA 52242, USA
[3] Department of Pediatrics, Carver College of Medicine at the University of Iowa, Iowa City, IA 52242, USA
* Correspondence: jordan-schultz@uiowa.edu; Tel.: +1-319-384-9388

Received: 24 July 2020; Accepted: 24 August 2020; Published: 26 August 2020

Abstract: Patients with adult-onset Huntington's Disease (AOHD) have been found to have dysfunction of the autonomic nervous system that is thought to be secondary to neurodegeneration causing dysfunction of the brain–heart axis. However, this relationship has not been investigated in patients with juvenile-onset HD (JOHD). The aim of this study was to compare simple physiologic measures between patients with JOHD ($n = 27$ participants with 64 visits) and participants without the gene expansion that causes HD (GNE group; $n = 259$ participants with 395 visits). Using data from the Kids-JOHD study, we compared mean resting heart rate (rHR), systolic blood pressure (SBP), and diastolic blood pressure (DBP) between the JOHD and GNE groups. We also divided the JOHD group into those with childhood-onset JOHD (motor diagnosis received before the age of 13, [$n = 16$]) and those with adolescent-onset JOHD (motor diagnosis received at or after the age of 13 [$n = 11$]). We used linear mixed-effects models to compare the group means while controlling for age, sex, and parental socioeconomic status and including a random effect per participant and family. For the primary analysis, we found that the JOHD group had significant increases in their rHR compared to the GNE group. Conversely, the JOHD group had significantly lower SBP compared to the GNE group. The JOHD group also had lower DBP compared to the GNE group, but the results did not reach significance. SBP and DBP decreased as disease duration of JOHD increased, but rHR did not continue to increase. Resting heart rate is more sensitive to changes in autonomic function as compared to SBP. Therefore, these results seem to indicate that early neurodegenerative changes of the central autonomic network likely lead to an increase in rHR while later progression of JOHD leads to changes in blood pressure. We hypothesize that these later changes in blood pressure are secondary to neurodegeneration in brainstem regions such as the medulla.

Keywords: juvenile-onset Huntington's Disease; autonomic; neurodegeneration

---

1. Introduction

Huntington's Disease (HD) is an inherited, neurodegenerative disease that causes motor, cognitive, and behavioral symptoms [1]. These symptoms are thought to be caused by striatal degeneration, which is the hallmark of HD [2,3]. However, there are myriad symptoms that can also occur as a result of the neurodegenerative processes associated with HD. Dysfunction of the autonomic nervous system (ANS) has been described in patients with adult-onset HD (AOHD) [4–8]. Specifically, patients with AOHD seem to have enhanced sympathetic tone compared to healthy controls. Structural and functional changes in the central autonomic network (CAN) of the brain are thought to drive autonomic dysfunction in HD [9,10]. Patients with juvenile-onset HD (JOHD) represent a rare group of individuals with significant neurodegeneration that begins very early in life, but physiologic measures of ANS function have never been reported in this patient population. Given the unique neurodegenerative changes that occur in JOHD, we would hypothesize that these patients would demonstrate some signs of a dysregulated ANS that occurs secondary to brain atrophy. We leveraged a large dataset of patients with JOHD to test the hypothesis that patients with JOHD have a dysregulated ANS with a propensity for enhanced sympathetic tone compared to a control group. This study has the potential to advance our understanding how dysfunction in the CAN may impact peripheral measures of ANS function in JOHD.

2. Materials and Methods

2.1. Participants

For these analyses, we utilized data from the Kids-HD and Kids-JOHD studies [11–14]. Both studies were longitudinal neuroimaging studies that ran in parallel. They recruited participants from around the country who were between the ages of 6 and 26 years old to the University of Iowa. The Kids-HD study recruited participants with a family history of HD (parent or grandparent with a known history of HD). These participants underwent genetic testing to determine whether they were gene carriers of the mutation that causes HD. The results of this genetic testing were not revealed to the participants, their family members, or the research staff. One research team member received the genetic results and anonymized them. This allowed for the ethical conduct of genetic testing in children. Additionally, the Kids-HD study recruited a cohort of healthy controls without a family history of HD. For the present analysis, all participants from the Kids-HD study with a CAG repeat length of less than 36 were included in the GNE group. The Kids-JOHD study, in comparison, recruited participants with a known diagnosis of JOHD based on molecular confirmation and a diagnosis provided by a neurologist. Participants with a CAG repeat length of 36 or above who were not symptomatic were excluded from the current analysis.

The JOHD group was first evaluated as a whole. We then split the group into those with childhood-onset JOHD and those with adolescent-onset JOHD. Childhood-onset JOHD is defined as an age of motor diagnosis that occurred before the age of 13, while adolescent-onset JOHD occurred between the ages of 13 and 21.

There were 27 participants in the JOHD group accounting for 64 visits, and there were 259 participants in the GNE group accounting for 395 visits. Among the JOHD group, three participants had five visits, two participants had four visits, five participants had three visits, nine participants had two visits, and eight participants only had one visit. In the GNE group, 12 participants had four visits, 21 participants had three visits, 58 participants had two visits, and 168 participants only had one visit. Among the JOHD group, 11 of the 27 participants were classified as having adolescent-onset JOHD and the other 16 were classified as having childhood-onset JOHD.

2.2. Statistical Analyses

For the primary outcome of interest, the physiologic measures of resting heart rate (rHR), systolic blood pressure (SBP), and diastolic blood pressure (DBP) were analyzed. These measures were collected in the Clinical Research Unit at the University of Iowa by trained professionals with equipment that is well maintained and calibrated regularly. We constructed linear mixed-effects regression models to compare the estimated mean differences in these measures between the GNE group and the JOHD group controlling for age, sex, the use of medications that may increase blood pressure, the use of medications that may decrease blood pressure and parental socioeconomic status as well as random effects per participant and per family to account for similarities among siblings. Because we were investigating children and young adults, the use of medications, such as stimulants, that may impact blood pressure were controlled for. Medications that were considered to increase blood pressure were stimulants, including amphetamine and methylphenidate-based products. Medications that were considered to lower blood pressure were carbidopa-levodopa, clonidine, guanfacine, and any anti-hypertensive medications. These models were run to first evaluate the JOHD group together, and then again to compare the childhood-onset JOHD and adolescent-onset JOHD groups to the GNE groups. We performed two unplanned sensitivity analyses. First, we repeated the primary analyses that compared rHR, SBP, and DBP between the JOHD participants and the control group after removing all participants visits from participants who were using medications that increased or decreased blood pressure. Next, we evaluated the relationship between duration of disease and rHR, SBP, and DBP among participants in the JOHD group only using linear mixed-effects regression analyses. Since these models only included participants with JOHD, we added CAG repeat length as a covariate in the models. The models were also adjusted for age, the use of medications that increase blood pressure, and the use of medications known to decrease blood pressure.

RStudio was used for all statistical analyses and a *p*-value of <0.05 was considered statistically significant.

3. Results

Primary Outcomes

Baseline demographics between the groups are outlined in Table 1.

Table 1. Baseline demographics by groups.

	JOHD Group	**Controls**	***p*-Value**
N (Visits)	27 (64)	259 (395)	NA
Female, % (*n*)	55.6 (15)	53.5 (138)	0.998
Age (years), Mean ± SD	15.89 ± 6.06	12.34 ± 3.76	<0.001
CAG Repeats, Mean ± SD	72.19 ± 14.18	20.29 ± 3.91	<0.001
Parental SES, % (*n*)			0.356
1	0.0 (0)	0.8 (2)	
2	48.0 (12)	58.0 (149)	
3	40.0 (10)	37.4 (96)	
4	8.0 (2)	3.1 (8)	
5	4.0 (1)	0.8 (2)	
Missing	*N* = 2	*N* = 2	
BP Increasing Meds, % (*n*)	11.1 (3)	2.7 (7)	0.087
BP Decreasing Meds, % (*n*)	14.8 (4)	1.2 (3)	<0.001
Disease Duration (years), Mean ± SD	3.38 ± 3.04	NA	NA

BP, blood pressure; CAG, cytosine-adenine-guanine; JOHD, juvenile-onset Huntington's Disease; SD, standard deviation; SES, socioeconomic status.

Participants in the JOHD group had significant elevations in their rHR compared to the GNE group. However, the mean SBP in the JOHD group was significantly decreased in the patients with JOHD compared to the GNE group. The mean DBP was lower in the JOHD group as well, but the results did not reach statistical significance (Table 2).

Table 2. Primary outcomes.

	JOHD Group	**Controls**	***p*-Value**
rHR, Mean ± SE	88.56 ± 2.36	78.50 ± 0.83	<0.0001
SBP, Mean ± SE	109.96 ± 1.99	115.99 ± 0.71	0.0053
DBP, Mean ± SE	61.88 ± 1.17	64.25 ± 0.42	0.060

DBP, diastolic blood pressure; JOHD, juvenile-onset Huntington's Disease; rHR, resting heart rate; SBP, systolic blood pressure; SE, standard error.

The mean rHR of participants with adolescent-onset JOHD (87.79 ± 3.72 bpm) was significantly elevated compared to the GNE group (78.50 ± 0.83 bpm; t = 2.44, p = 0.016). The childhood-onset

JOHD group had a mean rHR of 89.12 ± 3.11 bpm, which was also significantly elevated relative to the GNE group ($t = 3.27$, $p = 0.001$). The childhood-onset and adolescent-onset JOHD groups did not differ significantly from one another ($p = 0.786$) (Figure 1a). The mean SBP of participants with adolescent-onset JOHD was 111.79 ± 3.14 mmHg compared to 116.00 ± 0.71 in the GNE group ($t = -1.30$, $p = 0.195$). There was a significant difference between the childhood-onset JOHD group and the GNE groups, though. Specifically, the childhood-onset group's mean SBP was 108.66 ± 2.63 mmHg ($t = -2.66$, $p = 0.0084$). The difference between the childhood-onset and adolescent-onset groups was not statistically significant ($p = 0.451$) (Figure 1b). The mean DBP of the GNE group was 64.26 ± 0.43 mmHg. The adolescent-onset group's mean DBP was lower (63.11 ± 1.84 mmHg) compared to the GNE group, but not significantly so ($t = -0.609$, $p = 0.543$). However, the childhood-onset JOHD group's mean DBP was significantly lower (60.99 ± 1.55 mmHg) compared to the GNE group ($t = -2.01$, $p = 0.045$) (Figure 1c).

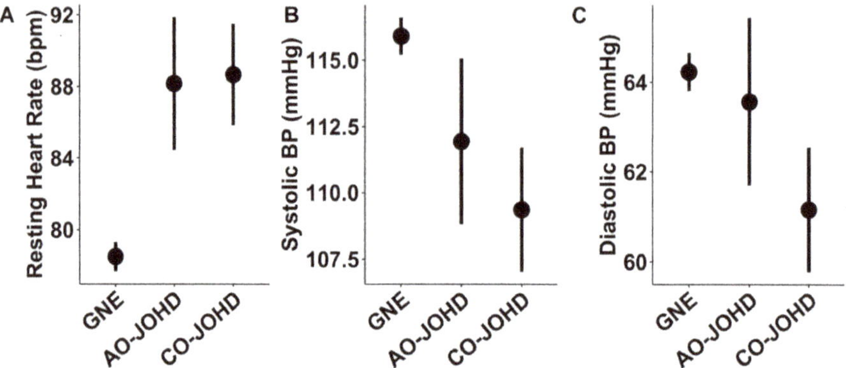

Figure 1. (**a**) Resting heart rate was significantly elevated in the adolescent-onset and childhood-onset JOHD groups compared to the GNE group. (**b**) Systolic BP was significantly decreased in the childhood-onset JOHD group compared to the GNE group. There was a gradual decrease in SBP from the GNE group to the adolescent-onset and childhood-onset JOHD groups. (**c**) Diastolic BP was significantly decreased in the childhood onset JOHD group compared to the GNE group. AO-JOHD: adolescent-onset JOHD; BP: blood pressure; Bpm: beats per minute; CO-JOHD: childhood-onset JOHD; GNE: gene-non-expanded group.

4. Discussion

Patients with HD produce a mutant form of the huntingtin protein. This protein is widely expressed in the brain, but also in peripheral tissues [15]. Consequently, peripheral manifestations of HD have been hypothesized to be caused by cellular dysfunction caused by expression of the mutant huntingtin protein in those tissues, independent of the known neurodegenerative process of HD. However, expression of the huntingtin protein is undetectable in the heart [15] and mouse models of HD have revealed changes to the cardiovascular system in the absence of mutant huntingtin aggregates in cardiac tissues, even at end stages of the disease [9]. Therefore, measured changes of physiologic markers of ANS function are more likely to be a secondary marker of neurodegeneration affecting central pathways in areas such as the CAN. Here, we have demonstrated for the first time that patients with JOHD have significant changes in physiologic measures of ANS function compared to healthy controls. Importantly, we hypothesize that these changes are a direct consequence of pathologic changes that occur in the central nervous system of patients with JOHD rather than peripheral manifestations of the disease.

Similar to previous reports of enhanced sympathetic tone in patients with AOHD, patients with JOHD had elevations in their rHR compared to healthy controls. However, the patients with JOHD had decreases in their SBP and DBP relative to the control group, which was unexpected. It is unclear what is driving the observed decrease in blood pressure in JOHD, but we hypothesize that earlier neurodegenerative changes may affect regions within the CAN that impact rHR. As patients progress through their disease, we believe that brain regions that are more closely related to blood pressure control are impacted. For example, the medulla oblongata receives afferent signals from the baroreceptors and affects blood pressure control [16]. The medulla has not been widely described as a primary area of neurodegeneration in HD using neuroimaging techniques. However, a post-mortem analysis demonstrated degeneration in areas of the brainstem in patients with HD [17]. This seems to fit with our hypothesis because neuroimaging studies are typically conducted in patients who have not reached the end stages of the disease, but post-mortem analyses obviously would mostly represent patients with end-stage disease, similar to patients with JOHD. To further investigate this theory, we conducted an unplanned analysis to investigate the relationship between disease duration and physiologic measures of cardiac function in the JOHD group. We found no significant relationship between disease duration and rHR ($p = 0.936$; Figure 2a). This supports the notion that rHR becomes significantly elevated early in the disease course of JOHD but does not seem to continue to worsen as the disease progresses. However, there were significant, negative correlations between disease duration and SBP ($t = -2.20$, $p = 0.037$; Figure 2b) and DBP ($t = -2.13$, $p = 0.044$; Figure 2c). These results further confirm our hypothesis that rHR is impacted earlier in the disease process of JOHD and changes in BP may be indicative of later-stage neurodegeneration.

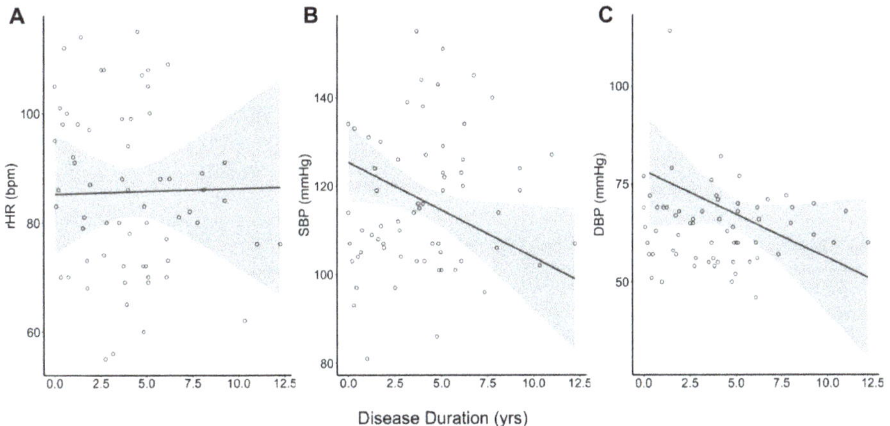

Figure 2. (**A**) Disease duration among patients with JOHD did not significantly predict rHR. (**B**) Systolic BP and (**C**) diastolic BP become significantly more decreased as disease duration increased in the JOHD group. Bpm: beats per minute; DBP: diastolic blood pressure; rHR: resting heart rate; SBP: systolic blood pressure.

This analysis is the first report (to the best of our knowledge) of physiologic markers of cardiac function being disrupted in JOHD. These results come from one of the largest datasets of patients with JOHD in the world. Despite this, there are important limitations to our work. As noted above, these results only allow us to report associations but are not meant to demonstrate a causative relationship between rHR and SBP in JOHD. Similarly, the Kids-JOHD and Kids-HD studies were not focused on cardiovascular measures. While all participants had their vital signs collected in a similar setting with similar equipment by trained medical professionals, confounding factors related to the collection of rHR and SBP could have been present and may have affected these results. Additionally, we recognize that rHR and SBP are surrogate measures of ANS function. Therefore, future studies focused on investigating ANS function in JOHD should collect more precise measures, such as heart rate variability or baroreflex sensitivity. Another limitation is the potential influence of medication use on these results. We have attempted to control for this confounder by including the use of medication that increase or decrease blood pressure as a covariate in all models. However, this may not have adequately controlled for the impact of medications. To further ensure that medication use was not significantly influencing these results, we performed an unplanned sensitivity analysis where we repeated the primary analyses in participants who were not taking a medication that is known to impact blood pressure. After doing this, the participants in the JOHD group still had a significantly elevated rHR and a significantly lower SBP compared to the GNE group. Interestingly, the JOHD group also had a significantly lower DBP compared to the GNe group in this analysis. While medication use is still an important confounder, this sensitivity analysis seems to indicate that the use of these medications is not significantly impacting the reported results. It is important to note that we hypothesize that the measured changes in rHR, SBP, and DBP in patients with JOHD is mediated by CNS alterations given that JOHD is a neurodegenerative disease. However, further research is needed to support this hypothesis. It is possible that the observed changes are mediated by the peripheral nervous system, the cardiovascular system (including physical fitness), and metabolic rate. Further studies are required to determine the root cause of autonomic dysfunction in JOHD. Lastly, it is important to note that the lower blood pressures in the JOHD group were at rest. Previous reports in patients with Parkinson's Disease and Multiple System Atrophy performed orthostatic tests to determine whether patients had a precipitous drop in their BP when going from lying down to standing up, which is more indicative of autonomic function. The present study did not perform any specific measures that were meant to perturb the autonomic nervous system. Therefore, the patterns identified at rest are only theorized to be associated with central changes that could affect the cardiovascular system.

5. Conclusions

Patients with JOHD seem to have elevated rHR compared to healthy controls. Additionally, the JOHD participants had significantly decreased SBP compared to the healthy controls. These changes seem to be indicative of progressive neurodegeneration and may further advance our understanding of the neurobiology of JOHD.

Author Contributions: Conceptualization, J.L.S. and P.C.N.; methodology, J.L.S. and P.C.N.; software, J.L.S.; validation, J.L.S. and P.C.N.; formal analysis, J.L.S.; investigation, J.L.S. and P.C.N.; resources, P.C.N.; data curation, P.C.N.; writing—original draft preparation, J.L.S.; writing—review and editing P.C.N.; visualization, J.L.S.; supervision, P.C.N.; project administration, P.C.N.; funding acquisition P.C.N. All authors have read and agreed to the published version of the manuscript.

Funding: This study was supported by National Institute of Neurological Disorders and Stroke (NINDS; R01NS055903). The APC was funded by National Institute of Neurological Disorders and Stroke (NINDS; R01NS055903).

Acknowledgments: We thank all of the co-investigators of the Kids-HD and Kids-JOHD study, as well as the participants and their families.

Conflicts of Interest: The authors declare no conflict of interest. The funders had no role in the design of the study; in the collection, analyses, or interpretation of data; in the writing of the manuscript, or in the decision to publish the results.

References

1.　MacDonald, M.E.; Ambrose, C.M.; Duyao, M.P.; Myers, R.H.; Lin, C.; Srinidhi, L.; Barnes, G.; Taylor, S.A.; James, M.; Groot, N.; et al. A novel gene containing a trinucleotide repeat that is expanded and unstable on Huntington's disease chromosomes. The Huntington's Disease Collaborative Research Group. *Cell* **1993**, *72*, 971–983. [CrossRef]

2. Tabrizi, S.J.; Scahill, R.I.; Owen, G.; Durr, A.; Leavitt, B.R.; Roos, R.A.; Borowsky, B.; Landwehrmeyer, B.; Frost, C.; Johnson, H.; et al. Predictors of phenotypic progression and disease onset in premanifest and early-stage Huntington's disease in the TRACK-HD study: Analysis of 36-month observational data. *Lancet Neurol.* **2013**, *12*, 637–649. [CrossRef]

3. Paulsen, J.S.; Long, J.D.; Ross, C.A.; Harrington, D.L.; Erwin, C.J.; Williams, J.K.; Westervelt, H.J.; Johnson, H.J.; Aylward, E.H.; Zhang, Y.; et al. Prediction of manifest Huntington's disease with clinical and imaging measures: A prospective observational study. *Lancet Neurol.* **2014**, *13*, 1193–1201. [CrossRef]

4. Andrich, J.; Schmitz, T.; Saft, C.; Postert, T.; Kraus, P.; Epplen, J.T.; Przuntek, H.; Agelink, M.W. Autonomic nervous system function in Huntington's disease. *J. Neurol. Neurosurg. Psychiatry* **2002**, *72*, 726–731. [CrossRef] [PubMed]

5. Den Heijer, J.C.; Bollen, W.L.; Reulen, J.P.; van Dijk, J.G.; Kramer, C.G.; Roos, R.A.; Buruma, O.J. Autonomic nervous function in Huntington's disease. *Arch. Neurol.* **1988**, *45*, 309–312. [CrossRef] [PubMed]

6. Kobal, J.; Meglic, B.; Mesec, A.; Peterlin, B. Early sympathetic hyperactivity in Huntington's disease. *Eur. J. Neurol.* **2004**, *11*, 842–848. [CrossRef] [PubMed]

7. Kobal, J.; Melik, Z.; Cankar, K.; Bajrovic, F.F.; Meglic, B.; Peterlin, B.; Zaletel, M. Autonomic dysfunction in presymptomatic and early symptomatic Huntington's disease. *Acta Neurol. Scand.* **2010**, *121*, 392–399. [CrossRef] [PubMed]

8. Kobal, J.; Melik, Z.; Cankar, K.; Strucl, M. Cognitive and autonomic dysfunction in presymptomatic and early Huntington's disease. *J. Neurol.* **2014**, *261*, 1119–1125. [CrossRef] [PubMed]

9. Mielcarek, M.; Inuabasi, L.; Bondulich, M.K.; Muller, T.; Osborne, G.F.; Franklin, S.A.; Smith, D.L.; Neueder, A.; Rosinski, J.; Rattray, I.; et al. Dysfunction of the CNS-heart axis in mouse models of Huntington's disease. *PLoS Genet.* **2014**, *10*, e1004550. [CrossRef]

10. Critchley, B.J.; Isalan, M.; Mielcarek, M. Neuro-Cardio Mechanisms in Huntington's Disease and Other Neurodegenerative Disorders. *Front. Physiol.* **2018**, *9*, 559. [CrossRef] [PubMed]

11. Moser, A.D.; Epping, E.; Espe-Pfeifer, P.; Martin, E.; Zhorne, L.; Mathews, K.; Nance, M.; Hudgell, D.; Quarrell, O.; Nopoulos, P. A survey-based study identifies common but unrecognized symptoms in a large series of juvenile Huntington's disease. *Neurodegener Dis. Manag.* **2017**, *7*, 307–315. [CrossRef] [PubMed]

12. Tereshchenko, A.; McHugh, M.; Lee, J.K.; Gonzalez-Alegre, P.; Crane, K.; Dawson, J.; Nopoulos, P. Abnormal Weight and Body Mass Index in Children with Juvenile Huntington's Disease. *J. Huntingt. Dis.* **2015**, *4*, 231–238. [CrossRef] [PubMed]

13. Tereshchenko, A.; Magnotta, V.; Epping, E.; Mathews, K.; Espe-Pfeifer, P.; Martin, E.; Dawson, J.; Duan, W.; Nopoulos, P. Brain structure in juvenile-onset Huntington disease. *Neurology* **2019**, *92*, e1939–e1947. [CrossRef] [PubMed]

14. van der Plas, E.; Langbehn, D.R.; Conrad, A.L.; Koscik, T.R.; Tereshchenko, A.; Epping, E.A.; Magnotta, V.A.; Nopoulos, P.C. Abnormal brain development in child and adolescent carriers of mutant huntingtin. *Neurology* **2019**, *93*, e1021–e1030. [CrossRef] [PubMed]

15. Carroll, J.B.; Bates, G.P.; Steffan, J.; Saft, C.; Tabrizi, S.J. Treating the whole body in Huntington's disease. *Lancet Neurol.* **2015**, *14*, 1135–1142. [CrossRef]

16. Colombari, E.; Sato, M.A.; Cravo, S.L.; Bergamaschi, C.T.; Campos, R.R., Jr.; Lopes, O.U. Role of the medulla oblongata in hypertension. *Hypertension* **2001**, *38*, 549–554. [CrossRef] [PubMed]

17. Rub, U.; Hentschel, M.; Stratmann, K.; Brunt, E.; Heinsen, H.; Seidel, K.; Bouzrou, M.; Auburger, G.; Paulson, H.; Vonsattel, J.P.; et al. Huntington's disease (HD): Degeneration of select nuclei, widespread occurrence of neuronal nuclear and axonal inclusions in the brainstem. *Brain Pathol.* **2014**, *24*, 247–260. [CrossRef] [PubMed]

Communication

The Association between CAG Repeat Length and Age of Onset of Juvenile-Onset Huntington's Disease

Jordan L. Schultz [1,2,*], Amelia D. Moser [3] and Peg C. Nopoulos [1,2,4]

[1] Department of Psychiatry, Carver College of Medicine at the University of Iowa, Iowa City, IA 52242, USA; peggy-nopoulos@uiowa.edu

[2] Department of Neurology, Carver College of Medicine at the University of Iowa, Iowa City, IA 52242, USA

[3] Department of Psychology and Neuroscience, University of Colorado Boulder, Boulder, CO 80309, USA; amelia.moser@colorado.edu

[4] Department of Pediatrics, Carver College of Medicine at the University of Iowa, Iowa City, IA 52242, USA

* Correspondence: jordan-schultz@uiowa.edu; Tel.: +1-319-384-9388

Received: 27 July 2020; Accepted: 18 August 2020; Published: 20 August 2020

Abstract: There is a known negative association between cytosine–adenine–guanine (CAG) repeat length and the age of motor onset (AMO) in adult-onset Huntington's Disease (AOHD). This relationship is less clear in patients with juvenile-onset Huntington's disease (JOHD), however, given the rarity of this patient population. The aim of this study was to investigate this relationship amongst a relatively large group of patients with JOHD using data from the Kids-JOHD study. Additionally, we analyzed data from the Enroll-HD platform and the Predict-HD study to compare the relationship between CAG repeat length and AMO amongst patients with AOHD to that amongst patients with JOHD using linear regression models. In line with previous reports, the variance in AMO that was predicted by CAG repeat length was 59% ($p < 0.0001$) in the Predict-HD study and 57% from the Enroll-HD platform ($p < 0.0001$). However, CAG repeat length predicted 84% of the variance in AMO amongst participants from the Kids-JOHD study ($p < 0.0001$). These results indicate that there may be a stronger relationship between CAG repeat length and AMO in patients with JOHD as compared to patients with AOHD. These results provide additional information that may help to model disease progression of JOHD, which is beneficial for the planning and implementation of future clinical trials.

Keywords: CAG; juvenile-onset Huntington's disease; motor onset

1. Introduction

Huntington's disease (HD) is an autosomal dominant, neurodegenerative disorder that causes cognitive, behavioral, and motor symptoms [1]. An abnormal number of repeats (≥ 36) of cytosine–adenine–guanine (CAG) within the huntingtin gene causes the mutation that leads to HD. The negative relationship between the number of CAG repeats that a person has and the age of motor onset (AMO) has been well-established for patients with adult-onset HD (AOHD) [2–6]. However, given the rarity of juvenile-onset HD (JOHD), the relationship between CAG repeat length and AMO is less clear. In general, it is believed that the correlation between CAG repeat length and AMO increases in higher CAG repeat lengths [7–9], but these reports are limited by small sample sizes. A previous review paper utilized data from 15 patients with JOHD who had a CAG repeat length of >60 to demonstrate that the association between AMO and CAG seemed to be higher at CAG lengths of 60 to 80. However, at CAG repeat lengths above approximately 80, the association between AMO and CAG became much weaker [10]. In order to further investigate this phenomenon in a larger group, the same authors gathered patient data from seven separate case studies or case series of patients with JOHD and found 26 patients with a CAG repeat length of >80. In this cohort, CAG repeat length accounted for only 26% of the variance in AMO [10]. A recent systematic review was performed and gathered information on more than 200 patients with reported JOHD [11]. This cohort of patients included a wide range of CAG repeats from 39 to 265. The authors reported a Pearson's correlation coefficient of −0.56 between CAG repeat length and the age of onset of clinical symptoms. In line with other studies, the strength of correlation seemed to weaken at longer CAG repeats [11]. Both of these large reviews are limited, however, by their reliance on a collection of data, rather than primary collection from a single cohort. Thus, the aim of the present study was to investigate the relationship between CAG repeat length and AMO amongst patients with JOHD, with the goal of providing additional clarity to this current gap in knowledge. We also leveraged data from the Enroll-HD platform and the Predict-HD study to compare the results of patients with JOHD to those of patients with AOHD.

2. Materials and Methods

2.1. Description of Data

We analyzed data from a large sample of participants enrolled in the Kids-JOHD studies ($N = 27$). These studies recruited participants from across the United States who were between the ages of 6 and 25 at the time of their first study visit and had a confirmed diagnosis of JOHD. To be enrolled, all participants were required to have already had molecular confirmation of HD and a clinical diagnosis of JOHD before traveling to the University of Iowa for assessment. To confirm that the participant was manifesting significant motor symptoms (meaning that they would meet criteria for a clinical diagnosis of JOHD), we utilized a combination of (1) detailed parental report of motor symptoms; (2) clinical evaluation by a pediatric neurologist at the University of Iowa; and (3) a Unified Huntington's Disease Ratings Scale (UHDRS) Total Motor Score (TMS) [12] of at least 12 to consider a participant to be motor-manifest. Each subject was required to have all three to be considered motor-manifest. Although a TMS of 12 may be considered low, patients with JOHD often present with far greater hypokinetic symptoms than do those with AOHD, and the UHDRS may not be sufficiently sensitive to these as it was designed for AOHD. These criteria excluded 10 individuals enrolled in the Kids-JOHD study and represented subjects who were tested locally mainly based on family history or behavioral issues, rather than significant motor symptoms. Of note, the TMS derived from the UHDRS takes into account various motor symptoms beyond just chorea. At their baseline visit, the participants with JOHD demonstrated the highest scores in saccade initiation, finger tapping, and rigidity (Figure 1). Additionally, the summative chorea score was one of the areas where patients with JOHD had the lowest score. This is of particular importance to consider when confirming the diagnosis of JOHD, as symptoms other than chorea can often be present. Participants were also excluded if they had a history of brain surgery or significant head trauma. All participants and their guardians (if under 18) signed informed consent prior to enrolling in these studies, which were approved by the University of Iowa Institutional Review Board (IRB).

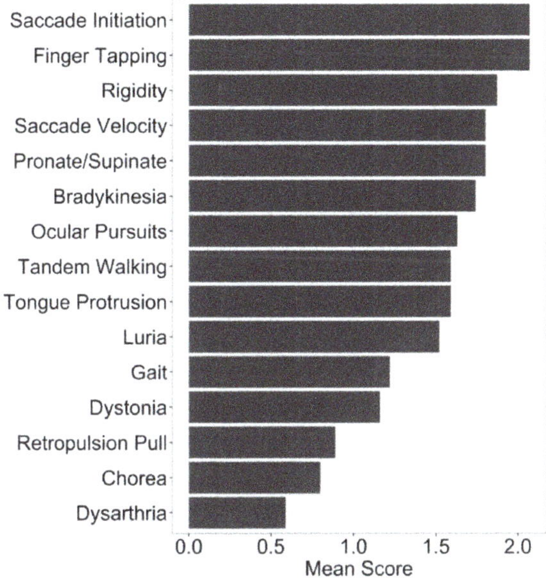

Figure 1. Mean scores on the various domains of the Unified Huntington's Disease Rating Scale (UHDRS). Specifically, ocular pursuit score is the mean of horizontal and vertical pursuit scores; saccade initiation is the mean of the horizontal and vertical initiation scores; saccade velocity is the mean of the horizontal and vertical velocity scores; finger tapping is the mean of the right and left hand scores; pronate/supinate is the mean of the right and left hand scores; rigidity is the mean of the scores in the right and left arms; dystonia score is the mean scores from the trunk, right and left upper extremities, and right and left lower extremities; chorea score is the mean of maximal chorea scores from the face, buccal-oral-lingual scores, trunk, right and left upper extremities, and right and left lower extremities. All other scores include only one component and represent the mean of that domain.

We also analyzed data from the Predict-HD study [13] and from the Enroll-HD platform [14]. Specifically, we included participants from both studies who received a motor diagnosis of HD during the study after the age of 21 and who had a CAG repeat length of less than 60. This was done to ensure that we were investigating only those participants with adult-onset HD (AOHD). Participant reporting of historical motor onset timing is included in the Enroll-HD platform, but this may be subject to significant recall bias. Participants must have had a motor exam conducted with a diagnostic confidence level (DCL) of less than four to be included in the analyses from Enroll-HD and Predict-HD. The age at which these participants had their first report of a diagnostic confidence level of four on the UHDRS [12] was considered the age of motor onset (AMO). There were 242 participants who received a motor diagnosis during the Predict-HD study and 782 participants in the Enroll-HD database.

2.2. Statistical Analysis

The primary analysis investigated the relationship between CAG repeat length and AMO amongst the JOHD participants using simple regression models. We used natural cubic splines to transform the independent variable (CAG repeat length) to investigate nonlinear relationships for the primary analysis. RStudio was used for all analyses, with a p-value of <0.05 considered statistically significant.

2.3. Institutional Ethical Approval

The University of Iowa institutional review board initially approved the study on 10/13/2011 (IRB # 201109879). For participants younger than 18 years, parents or guardians provided written consent and children provided written assent. For participants who were 18 years or older, participants provided written consent.

3. Results

The participants from the Kids-JOHD study had a significant nonlinear relationship between AMO and CAG repeat length ($R^2 = 0.84$, $p = 2.63 \times 10^{-10}$) (Figure 2A). As has been reported previously, we also observed significant nonlinear relationships between AMO and CAG repeat length using data from participants with AOHD from the Predict-HD and Enroll-HD studies. From Predict-HD, the R^2 value was 0.59 ($p = 2.2 \times 10^{-16}$) (Figure 2B). The results from participants from Enroll-HD closely matched those observed in Predict-HD. Specifically, the R^2 value was 0.57 ($p = 2.2 \times 10^{-16}$) (Figure 2C).

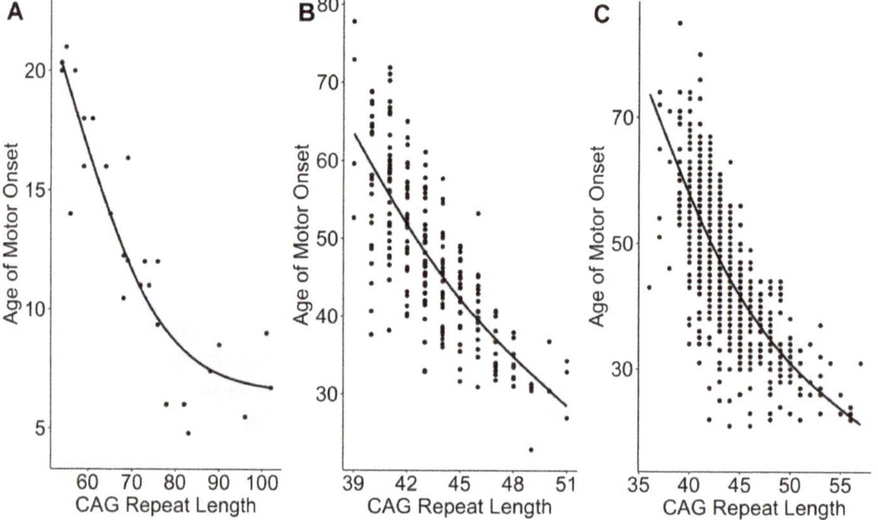

Figure 2. Cytosine–adenine–guanine (CAG) repeat length significantly predicts the age of motor onset in (**A**) patients with juvenile-onset Huntington's disease (JOHD) from the Kids-JOHD study, (**B**) patients with adult-onset Huntington's Disease (AOHD) from the Predict-HD study, and (**C**) patients with AOHD from the Enroll-HD study. Black lines show the predicted regression lines, and the gray ribbons display the 95% confidence intervals.

4. Discussion

In the present analysis, we demonstrated that the CAG repeat length accounted for over 80% of the variance in AMO amongst patients with JOHD. This was substantially higher than in either of the AOHD analyses conducted, which showed CAG repeat length accounting for 59% of the variance (Predict-HD sample) and 57% of the variance (Enroll-HD). These results support previous reports demonstrating potentially increased predictive power of higher CAG repeat lengths [7–9]. Of note, previous reports have demonstrated that the predictive power of CAG repeat length on AMO seems to decrease at the highest CAG repeats of approximately 80 or above [10,11]. Our current cohort only included seven participants with a CAG repeat length above 80. Therefore, we did not have sufficient data to formally analyze whether or not the relationship between CAG and AMO weakens at higher CAG repeat lengths. However, informally, our results seem to confirm these previous reports. In Figure 2A, it seems as though there is a strong, linear relationship between CAG repeat length and AMO in those participants with a CAG repeat length of <80. There seems to be a bend in the regression line at approximately a CAG repeat length of 80, where the line begins to flatten out. This same shape was seen in previous reports with larger numbers of patients [10,11]. This may be due to a floor effect in the ability of the CAG repeat length to predict AMO. Specifically, it may be possible that neurodegenerative changes occur over the course of 3–5 years. Therefore, the earliest possible AMO may be approximately five years old, regardless of CAG repeat length. This is only a hypothesis, though. Another possible explanation for the weakened relationship between CAG and AMO at CAG repeats above 80 may be related to the role that the huntingtin protein plays in neurodevelopment [15,16]. Neurodevelopmental changes have been reported to be more prominent at higher CAG repeats in patients with AOHD [15]. At higher CAG repeats, it is possible that the neurodevelopmental aberrations play a major role in the AMO in addition to neurodegeneration. This likely leads to significant difficulty in determining when their actual AMO is, likely resulting in a significant amount of heterogeneity in age of diagnosis.

One possible explanation for why higher CAG repeat lengths (>60) may explain more of the variance in AMO is that longer CAG repeat lengths may play a greater role in the development of pathologic changes that impact the onset of disease. Another consideration is that it is known that environmental exposures may modify disease onset in patients with AOHD [17–19]. Patients with JOHD may not have the same opportunity to be exposed to particular environmental factors. Therefore, their AMO is more closely linked to CAG repeat length alone and not the additional impact of environmental factors. Genetic modifiers of disease onset have also been identified in patients with AOHD [20,21]. These studies rely on large numbers of patients to identify genetic modifiers that may impact disease onset in AOHD. It cannot be ruled out that specific genetic modifiers exist that could impact the AMO in JOHD that have not been identified, given the rarity of this patient population.

One of the largest previous reports of the relationship between CAG repeat length and AMO amongst patients with JOHD utilized data from the Italian Huntington's Disease Databank, which retrospectively collected data from patients at two separate institutions and only includes 15 patients with a CAG repeat length of >60 [10,22]. This same review gathered data from seven separate case reports and case series and identified 26 patients with more than 80 CAG repeats [10]. Given the means by which these data were collected, the conclusions drawn in the resulting review may be seen as preliminary, as they are not the result of primary data collection. However, using the present large dataset of patients with JOHD, we are now able to confirm these previous findings showing that the strength of correlation between CAG and AMO is greater in JOHD, predicting about 84% of the variance in AMO [10].

There are important limitations to this study. First, despite being one of the largest studies of JOHD in the world, the number of patients is still relatively small. Second, we did not implement natural logarithmic transformations of our data, which has been recommended in previous studies investigating CAG and AMO [23]. We opted to not use natural logarithmic transformations of the data because doing so appeared to lead to disproportionate variance across the groups, which would lead to heteroscedasticity in the data. Lastly, as mentioned previously, the diagnosis of motor symptoms in children with the longest CAG repeats can be quite difficult and subject to bias and variability.

5. Conclusions

This study has provided further confirmation that CAG repeat length significantly predicts AMO in patients with JOHD. In fact, more than 80% of the variance of AMO was explained by CAG repeat length. This provides additional information that allows for more accurate modeling of JOHD, which is critical for future clinical trials.

Author Contributions: Conceptualization, J.L.S., A.D.M., and P.C.N.; methodology, J.L.S. and P.C.N.; software, J.L.S.; validation, A.D.M. and P.C.N.; formal analysis, J.L.S.; investigation, J.L.S. and A.D.M.; resources, P.C.N.; data curation, P.C.N.; writing—original draft preparation, J.L.S.; writing—review and editing, A.D.M. and P.C.N.; visualization, J.L.S.; supervision, P.C.N.; project administration, P.C.N.; funding acquisition, P.C.N. All authors have read and agreed to the published version of the manuscript.

Funding: This study was supported by National Institute of Neurological Disorders and Stroke (NINDS; R01NS055903). The APC was funded by National Institute of Neurological Disorders and Stroke (NINDS; R01NS055903).

Acknowledgments: Enroll-HD is a longitudinal observational study for Huntington's disease families intended to accelerate progress towards therapeutics; it is sponsored by CHDI Foundation, a nonprofit biomedical research organization exclusively dedicated to developing therapeutics for HD. Enroll-HD would not be possible without the vital contributions of the research participants and their families. We also thank the patients and families that participated in the Predict-HD study and the Kids-JOHD study. We also wish to extend our thanks to the CHDI Foundation for their dedication to advancing scientific discovery in the HD community through data sharing.

Conflicts of Interest: The authors declare no conflict of interest. The funders had no role in the design of the study; in the collection, analyses, or interpretation of data; in the writing of the manuscript; or in the decision to publish the results.

References

1. Macdonald, M.E.; Ambrose, C.M.; Duyao, M.P.; Myers, R.H.; Lin, C.; Srinidhi, L.; Barnes, G.; Taylor, S.A.; James, M.; Groot, N.; et al. A novel gene containing a trinucleotide repeat that is expanded and unstable on Huntington's disease chromosomes. *Cell* **1993**, *72*, 971–983. [CrossRef]

2. Langbehn, D.R.; Brinkman, R.; Falush, D.; Paulsen, J.S.; Hayden, M.R. A new model for prediction of the age of onset and penetrance for Huntington's disease based on CAG length. *Clin. Genet.* **2004**, *65*, 267–277. [CrossRef]

3. Stine, O.; Pleasant, N.; Franz, M.L.; Abbott, M.H.; Folstein, S.E.; Ross, C.A. Correlation between the onset age of Huntington's disease and length of the trinucleotide repeat in IT-15. *Hum. Mol. Genet.* **1993**, *2*, 1547–1549. [CrossRef]

4. Lucotte, G.; Turpin, J.; Riess, O.; Epplen, J.T.; Siedlaczk, I.; Loirat, F.; Hazout, S. Confidence intervals for predicted age of onset, given the size of (CAG) n repeat, in Huntington's disease. *Qual. Life Res.* **1995**, *95*, 231–232. [CrossRef]

5. Andrew, S.E.; Goldberg, Y.P.; Kremer, B.; Telenius, H.; Theilmann, J.; Adam, S.; Starr, E.; Squitieri, F.; Lin, B.; Kalchman, M.A.; et al. The relationship between trinucleotide (CAG) repeat length and clinical features of Huntington's disease. *Nat. Genet.* **1993**, *4*, 398–403. [CrossRef]

6. Squitieri, F.; Sabbadini, G.; Mandich, P.; Gellera, C.; Di Maria, E.; Bellone, E.; Castellotti, B.; Nargi, E.; De Grazia, U.; Frontali, M.; et al. Family and molecular data for a fine analysis of age at onset in Huntington disease. *Am. J. Med Genet.* **2000**, *95*, 366–373. [CrossRef]

7. Metzger, S.; Rong, J.; Nguyen, H.H.P.; Cape, A.; Tomiuk, J.; Soehn, A.S.; Freudenberg-Hua, Y.; Propping, P.; Freudenberg, J.; Tong, L.; et al. Huntingtin-associated protein-1 is a modifier of the age-at-onset of Huntington's disease. *Hum. Mol. Genet.* **2008**, *17*, 1137–1146. [CrossRef]

8. Sun, Y.-M.; Zhang, Y.-B.; Wu, Z.-Y. Huntington's Disease: Relationship Between Phenotype and Genotype. *Mol. Neurobiol.* **2016**, *54*, 342–348. [CrossRef]

9. Chao, T.-K.; Hu, J.; Pringsheim, T. Risk factors for the onset and progression of Huntington disease. *NeuroToxicology* **2017**, *61*, 79–99. [CrossRef]

10. Squitieri, F.; Frati, L.; Ciarmiello, A.; Lastoria, S.; Quarrell, O. Juvenile Huntington's disease: Does a dosage-effect pathogenic mechanism differ from the classical adult disease? *Mech. Ageing Dev.* **2006**, *127*, 208–212. [CrossRef]

11. Cronin, T.; Rosser, A.; Massey, T.H. Clinical Presentation and Features of Juvenile-Onset Huntington's Disease: A Systematic Review. *J. Huntington's Dis.* **2019**, *8*, 171–179. [CrossRef] [PubMed]

12. Kieburtz, K.; Penney, J.B.; Corno, P.; Ranen, N.; Shoulson, I.; Feigin, A.; Abwender, D.; Greenarnyre, J.T.; Higgins, D.; Marshall, F.J.; et al. Unified Huntington's disease rating scale: Reliability and consistency. *Mov. Disord.* **1996**, *11*, 136–142. [CrossRef]

13. Paulsen, J.S.; Long, J.D.; Rosser, A.E.; Harrington, D.L.; Erwin, C.J.; Williams, J.K.; Westervelt, H.J.; Johnson, H.J.; Aylward, E.H.; Zhang, Y.; et al. Prediction of manifest Huntington's disease with clinical and imaging measures: A prospective observational study. *Lancet Neurol.* **2014**, *13*, 1193–1201. [CrossRef]

14. Landwehrmeyer, G.B.; Fitzer-Attas, C.J.; Giuliano, J.D.; Gonçalves, N.; Anderson, K.E.; Cardoso, F.; Ferreira, J.J.; Mestre, T.A.; Stout, J.C.; Sampaio, C. Data Analytics from Enroll-HD, a Global Clinical Research Platform for Huntington's Disease. *Mov. Disord. Clin. Pr.* **2016**, *4*, 212–224. [CrossRef] [PubMed]

15. Van Der Plas, E.; Langbehn, D.R.; Conrad, A.L.; Koscik, T.R.; Tereshchenko, A.; Epping, E.; Magnotta, V.; Nopoulos, P. Abnormal brain development in child and adolescent carriers of mutant huntingtin. *Neurol.* **2019**, *93*, 1021–1030. [CrossRef]

16. Saudou, F.; Humbert, S. The Biology of Huntingtin. *Neuron* **2016**, *89*, 910–926. [CrossRef]
17. Schultz, J.; Kamholz, J.A.; Moser, D.J.; Feely, S.M.; Paulsen, J.S.; Nopoulos, P. Substance abuse may hasten motor onset of Huntington disease. *Neurol.* **2017**, *88*, 909–915. [CrossRef]
18. Schultz, J.L.; Nopoulos, P.C.; Killoran, A.; Kamholz, J.A. Statin use and delayed onset of Huntington's disease. *Mov. Disord.* **2018**, *34*, 281–285. [CrossRef]
19. Schultz, J.; Harshman, L.A.; Langbehn, D.R.; Nopoulos, P.C. Hypertension Is Associated with an Earlier Age of Onset of Huntington's Disease. *Mov. Disord.* **2020**. [CrossRef]
20. Lee, J.-M.; Wheeler, V.C.; Chao, M.J.; Vonsattel, J.P.G.; Pinto, R.M.; Lucente, D.; Abu-Elneel, K.; Ramos, E.M.; Mysore, J.S.; Gillis, T.; et al. Identification of Genetic Factors that Modify Clinical Onset of Huntington's Disease. *Cell* **2015**, *162*, 516–526. [CrossRef]
21. Lee, J.-M.; Correia, K.; Loupe, J.; Kim, K.-H.; Barker, D.; Hong, E.P.; Chao, M.J.; Long, J.D.; Lucente, D.; Vonsattel, J.P.G.; et al. CAG Repeat Not Polyglutamine Length Determines Timing of Huntington's Disease Onset. *Cell* **2019**, *178*, 887–900.e4. [CrossRef] [PubMed]
22. Squitieri, F.; Berardelli, A.; Nargi, E.; Castellotti, B.; Mariotti, C.; Cannella, M.; Lavitrano, M.L.; De Grazia, U.; Gellera, C.; Ruggieri, S. Atypical movement disorders in the early stages of Huntington's disease: Clinical and genetic analysis. *Clin. Genet.* **2000**, *58*, 50–56. [CrossRef] [PubMed]
23. Langbehn, D.R.; Hayden, M.R.; Paulsen, J.S.; The PREDICT-HD Investigators of the Huntington Study Group. CAG-repeat length and the age of onset in Huntington disease (HD): A review and validation study of statistical approaches. *Am. J. Med Genet. Part B Neuropsychiatr. Genet.* **2009**, *153B*, 397–408. [CrossRef] [PubMed]

Article

Behavioral Deficits in Juvenile Onset Huntington's Disease

Kathleen E. Langbehn [1],*, Ashley M. Cochran [1], Ellen van der Plas [1], Amy L. Conrad [2], Eric Epping [1], Erin Martin [1], Patricia Espe-Pfeifer [1] and Peg Nopoulos [1,2,3],*

[1] Department of Psychiatry, University of Iowa Carver College of Medicine, Iowa City, IA 52242, USA; ashley-cochran@uiowa.edu (A.M.C.); ellen-vanderplas@uiowa.edu (E.v.d.P.); eric-epping@uiowa.edu (E.E.); erin-martin@uiowa.edu (E.M.); patricia-espepfeifer@uiowa.edu (P.E.-P.)

[2] Stead Family Department of Pediatrics, University of Iowa Carver College of Medicine, Iowa City, IA 52242, USA; amy-l-conrad@uiowa.edu

[3] Department of Neurology, University of Iowa Carver College of Medicine, Iowa City, IA 52242, USA

* Correspondence: kathleen-langbehn@uiowa.edu (K.E.L.); peggy-nopoulos@uiowa.edu (P.N.); Tel.: +1-319-356-1144 (P.N.)

Received: 14 July 2020; Accepted: 7 August 2020; Published: 11 August 2020

Abstract: Reports of behavioral disturbance in Juvenile-Onset Huntington's Disease (JOHD) have been based primarily on qualitative caregiver reports or retrospective medical record reviews. This study aims to quantify differences in behavior in patients with JOHD using informant- and self-report questionnaires. Informants of 21 children/young adults (12 female) with JOHD and 115 children/young adults (64 female) with a family history of Huntington's Disease, but who did not inherit the disease themselves (Gene-Non-Expanded; GNE) completed the Behavior Rating Inventory of Executive Function (BRIEF) and the Pediatric Behavior Scale (PBS). Mixed linear regression models (age/sex adjusted) were conducted to assess group differences on these measures. The JOHD group had significantly higher scores, indicating more problems, than the GNE group on all BRIEF subscales, and measures of Aggression/Opposition and Hyperactivity/Inattention of the PBS (all $p < 0.05$). There were no group differences in Depression/Anxiety. Inhibit, Plan/Organize, Initiate, and Aggression/Opposition had significant negative correlations with Cytosine-Adenine-Guanine (CAG) repeat length (all $p < 0.05$) meaning that individuals with higher CAG repeats scored lower on these measures. There was greater discrepancy between higher informant-vs. lower self-reported scores in the JOHD group, supporting the notion of lack of insight for the JOHD-affected group. These results provide quantitative evidence of behavioral characteristics of JOHD.

Keywords: Huntington's disease; behavioral regulation; executive function; trinucleotide repeat disorder

---

1. Introduction

Huntington's disease (HD) is an autosomal dominant neurodegenerative disease caused by a Cytosine-Adenine-Guanine (CAG) repeat expansion of the *Huntingtin* gene (*HTT*). The disease typically presents in adulthood with an average age of onset of 40 (referred to as Adult Onset HD or AOHD), and is marked by a combination of motor, cognitive, and behavioral symptoms. Approximately 1–10% of patients with HD will experience onset of symptoms before age 21, which is categorized as Juvenile-Onset HD (JOHD) [1].

Evidence of behavioral disturbances in JOHD is primarily predicated on retrospective medical record analyses [2,3] and caregiver reports [4–6] with few attempts to systematically evaluate these changes prospectively. Behavioral problems and cognitive decline are often among the first symptoms to present in individuals with JOHD [6–9] and can emerge years before the onset of motor symptoms [5], a pattern that is parallel with that of AOHD. Behavioral issues reported in JOHD include violence, aggression, oppositional behavior, obsession, depression, anxiety, impulsivity, attention issues, psychosis, and substance abuse [1–4,8,10,11]. Family members and caregivers of individuals with JOHD have reported that behavioral symptoms are often more distressing and disruptive than motor symptoms [6].

The Kids-HD and Kids-JOHD study are parallel programs at the University of Iowa. The Kids-HD study enrolls children/young adults (ages 5–26 years old) who are at-risk for HD based on a parent or grandparent having been diagnosed with HD. These children are genotyped for research purposes only, and categorized into the Gene-Expanded (GE, CAG > 36) or Gene Non-Expanded (GNE, CAG < 35) group. Those that are GE will go on to develop AOHD later in life and those that are GNE will never develop HD. The Kids-JOHD study enrolls children/young adults (ages 5–26 years old) who are already symptomatic with JOHD and have had molecular confirmation of the gene expansion (motor diagnosis made prior to age 21). The GNE group makes an excellent comparison group for the JOHD sample given that although they did not inherit the gene, they are from a family in which a parent, and possibly other family members, are suffering from HD, a family environment similar to the JOHD participants. The first aim of the current study was to establish group differences between JOHD and GNE participants based on informant ratings of behavior. The second aim was to examine a potential CAG repeat length effect in the JOHD participants. The third aim was to explore differences between self-reported and informant-reported executive function in participants over the age of 18.

2. Materials and Methods

2.1. Participants

This sample consists of children and young adults (ages 5–26) who participated in the Kids-HD or Kids-JOHD studies at the University of Iowa. Details of the Kids-HD program can be found in the 2019 publication on brain development in the Kids-HD sample [12]. From the children/young adults at risk from the Kids-HD study, we utilized the Gene-Non-Expanded (GNE) as controls.

Recruitment for the Kids-JOHD study was done through the Center of Excellence at the University of Iowa and national Huntington's Disease Society of America events. To be eligible, participants required a genetic confirmation of an expanded CAG repeat. If the participant was older than 21 at the time of assessment, they were required to have had a clinical diagnosis prior to the age of 21.

All participants were recruited from across the United States and travelled to the University of Iowa to complete the study. The study followed an accelerated longitudinal design (ALD), where some individuals were assessed once, while others were assessed on multiple occasions with variable lengths of follow-up [12]. This design is less affected by attrition, which is especially important in this population due to disease progression. It also allows us to study the disease over a larger age range, and thus a greater number of participants. However, each age range may not be equally represented [13]. Participants completed multiple visits if they were willing and able to return for follow-up.

Both study protocols were approved by the Institutional Review Board (IRB) at the University of Iowa and were conducted in accordance with the Declaration of Helsinki. Parents or legally authorized representatives provided written consent for participants under the age of 18 or those who were unable to provide consent due to disease progression. Participants 18 years of age or older with the capacity to consent provided written informed consent for participation. This project was initially approved by the University of Iowa Institutional Review Board (IRB Number: 201109879) on 9 January 2012, and most recently approved on 15 May 2020.

2.2. Genetic Analysis

A DNA sample of either blood or saliva was obtained from participants in the Kids-HD study. The presence or absence of the mutant CAG expansion was determined using PCR analysis by the University of Iowa Molecular Diagnostic Laboratory. This analysis was done for research purposes only, and results were not disclosed to anyone including the participants, participants' families, or the clinical research study staff [12]. All the participants in the Kids-JOHD study had to have molecular confirmation prior to enrollment and came with medical record documentation of the gene expansion.

2.3. Motor Rating

Like AOHD, the diagnosis of JOHD requires the presence of significant motor abnormality. A few of the participants in the JOHD study had been tested locally, yet examination by the neurologist locally showed no significant, or only subtle, motor findings. Our aim was to examine a homogenous group of motor-manifest patients. To that end, all participants were assessed by a trained motor examiner using the Unified Huntington's Disease Rating Scale (UHDRS). A total motor score (TMS) was obtained by summing the core UHDRS items. To be included in the analysis, the JOHD participants were required to have a TMS of greater than 18. The rationale for the relatively high cut-off for the TMS is because the UHDRS is sensitive to developmental motor changes such that normal developing younger children will show higher scores than older children. Therefore, even in a large cohort of children at risk, but who did not inherit the gene expansion (Gene Non-Expanded or GNE), the UHDRS can be as high as 17, as shown by our previous analysis of the Kids-HD cohort [14]. In the current cohort, the highest TMS was 17 for a GNE participant. This was used as a guide for the JOHD group where the cut-off was determined to be 18 (with lowest TMS in the JOHD group being 19).

2.4. Behavioral Measures

Behavior Rating Inventory of Executive Function (BRIEF). While some standardized tests are designed to measure specific executive functioning skills in an individual, the BRIEF is a questionnaire designed to assess executive function behaviors [14]. Items are coded as "never", "sometimes", or "often" being a problem. The total Global Executive Composite score is divided into the Behavioral Regulation Index (BRI; Inhibit, Shift, Emotional Control) and the Metacognition Index (MI; Initiate, Working Memory, Plan/Organize, Organization of Materials, Monitor). Higher scores indicate more problematic behaviors.

If the participant was younger than 18 years old, the informant completed the BRIEF-Parent form; however, if the participant was 18 years old or older, the informant completed the BRIEF-Adult (BRIEF-A) Informant form. Informants were individuals who accompanied the participant to the research appointment. For participants younger than 18 years of age, the informant was a parent or legal guardian. For participants 18 years of age or older, the informant was a parent, guardian, spouse/partner, or friend. Additionally, participants age 18 and older completed the BRIEF-A Self Report Form.

With regard to the Pediatric Behavior Scale-30 (PBS-30), the PBS is a parent/informant-report measure designed to assess broad domains of functioning; it was derived from the full 165-item version of the PBS [15]. The PBS-30 includes 30 Likert-scale items ("almost never or not at all" to "very often or very much") that describe different behaviors, where higher scores reflect more problems. Four scales are calculated: Aggression/Opposition, Hyperactivity/Inattention, Depression/Anxiety, and Physical Health. The same form was used for all ages in the current study.

Supplementary Tables S1 and S2 summarize the number of observations available for each BRIEF and PBS subscale. Supplementary Table S3 summarizes the differences in scores between informant and self-reports for the BRIEF-A.

2.5. Statistical Analyses

Raw scores from the two scales were analyzed across the two groups via mixed linear regression models with individual BRIEF and PBS subscales as the outcome variables. Group, age, and sex were included as main effects in all regression models, and participant ID and family ID were included as random effects to account for non-independency of the observations. Including family ID in the model controlled for the effects of having more than one person from a family. Adding the participant ID controls meant that the correlation between repeat visits from a single individual was controlled for in the model. Sex by group interactions were entered into the model and subsequently removed if not significant.

The impact of CAG repeat length was examined in the JOHD group for measures that were significantly different between groups using mixed linear regression models to predict behavioral and executive functioning outcomes by CAG repeat expansion length.

Differences in BRIEF-A informant and self-reported scores were analyzed in patients 18 and older by calculating the difference in informant-reported and self-reported raw scores for each BRIEF-A subscale outcome measure. Difference scores were predicted with mixed linear effects models including main effects of group, age, and sex and controlling random effects of family ID. The difference score represents a difference in perspective on behavioral/cognitive problems between the participant and their parent. Thus, difference scores close to 0 represent agreement between self and proxy assessments, while large differences represent incongruent perspectives. All models were corrected for multiple comparisons using the False Discovery Rate (FDR) method. All analyses were completed using RStudio version 1.2.5042 (RStudio, PBC, Boston, MA, USA).

3. Results

3.1. Sample

The sample included 21 JOHD individuals (12 female). From this group, nine participants were seen once, nine were seen twice, two were seen three times, one was seen four times, and three were examined on five occasions, for a total of 49 observations.

There were 115 GNE individuals (64 female). From this group, 60 participants were seen once, 30 were seen twice, 17 were seen three times, and 8 were seen for 4 visits for a total of 203 observations. There were no significant differences in distribution of sex between JOHD and GNE individuals (χ^2 $(1, N = 137) = 1.10 \times 10^{-31}$, $p = > 0.99$). Average elapsed time between follow-up visits was 1.3 years (SD = 1.8 years). All participants were seen between March 2006 and February 2020 with an average of 1.65 visits and median of one visit.

Mean age at evaluation was significantly different between groups: JOHD patients were 15.23 years old on average (SD = 5.55) and GNE individuals were 13.47 years old on average (SD = 3.87 years; t (126.66) = 2.03), $p = 0.04$). CAG repeats ranged from 15 to 34 in the GNE group (median = 19) and from 54 to 102 in the JOHD group (median = 76). Distribution of total motor impairment scores (sum of core UHDRS items) ranged from 19 to 103 in the JOHD group. Average disease duration (defined as age at time of assessment minus age at time of clinical diagnosis) for individuals with JOHD was 3.6 years (SD = 1.5 years), meaning that most JOHD participants were early in the course of the motor manifest stage of the disease. Full group statistics are shown in Table 1.

Table 1. Demographics by groups.

	GNE (*N* = 203)	JOHD (*N* = 49)
Age		
Mean (SD)	13.5 (3.87)	15.2 (5.55)
Median (Min, Max)	13.7 (6.00, 22.5)	15.8 (5.08, 25.1)
Sex		
Females	120 (59.1%)	29 (59.2%)
Males	83 (40.9%)	20 (40.8%)
CAG		
Median (Min, Max)	19.0 (15.0, 34.0)	76.0 (54.0, 102)
TMS		
Mean (SD)	1.28 (2.67)	58.0 (21.3)
Median (Min, Max)	0 (0, 17.0)	55.0 (19.0, 103)

Note: GNE, Gene-Non-Expanded, i.e., participants with a family history of Huntington's Disease who did not inherit the mutant expansion; JOHD, participants with juvenile onset Huntington's Disease; CAG, CAG repeat expansion length; TMS, total motor score, calculated as cumulative Unified Huntington's Disease Rating (UHDRS) items.

3.2. Behavioral Performance Group Differences

The JOHD group had statistically significantly higher scores than the GNE group on all subscales of the BRIEF (Emotional Control, Inhibit, Shift, Monitor, Plan/Organize, and Working Memory all FDR < 0.001; Organization of Materials FDR = 0.0015; see Figure 1 and Table 2).

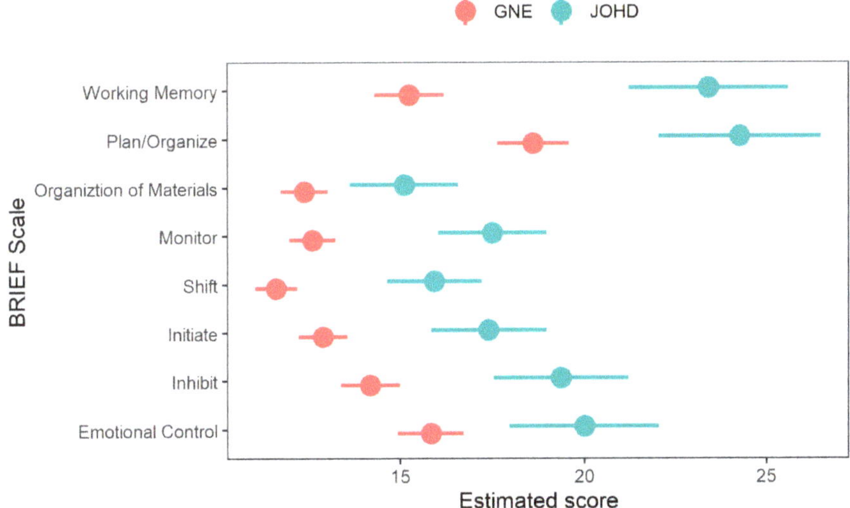

Figure 1. Differences in Behavior Rating Inventory of Executive Function (BRIEF) subscale scores between JOHD (blue) and GNE (red) participants. The x-axis shows age- and sex-adjusted estimates from mixed linear effects models after controlling for random effects of repeated measures and family ID. The y-axis shows subscales of the Behavioral Rating Inventory of Executive Function (BRIEF). The larger circles represent the means and horizontal lines indicate 95% confidence limits.

The JOHD group had statistically significantly higher scores than the GNE group on the Aggression/Opposition and Hyperactivity/Inattention subscales of the PBS (FDR = 0.00017 and <0.0001 respectively; see Figure 2 and Table 2). In contrast, there were no significant group differences in parent reported measures of Depression/Anxiety and Physical Health subscales of the PBS (both FDR > 0.1). There was no significant main effect of sex for any measure.

Table 2. BRIEF and PBS model statistics and marginal means.

Variable	Diff Means	*t*-Value (df)	FDR	Marg Mean GNE	95% CI GNE	Marg Mean JOHD	95% CI JOHD
Emotional Control	−4.19	$t_{(120)} = -3.68$	0.000466 *	15.83	14.93: 16.73	20.02	17.96: 22.08
Inhibit	−5.18	$t_{(120)} = -5.1$	2.62×10^{-6} *	14.19	13.39: 15.00	19.37	17.53: 21.21
Initiate	−4.5	$t_{(117)} = -5.16$	2.43×10^{-6} *	12.91	12.23: 13.58	17.41	15.82: 19.00
Shift	−4.31	$t_{(110)} = -5.99$	1.19×10^{-7} *	11.62	11.04: 12.19	15.93	14.63: 17.23
Self-Monitor	−4.89	$t_{(112)} = -5.96$	1.19×10^{-7} *	12.62	11.98: 13.26	17.51	16.02: 19.00
Organization of Materials	−2.72	$t_{(115)} = -3.32$	0.00147 *	12.40	11.75: 13.04	15.12	13.63: 16.60
Plan/Organize	−5.65	$t_{(115)} = -4.58$	2.02×10^{-5} *	18.62	17.64: 19.60	24.27	22.04: 26.50
Working Memory	−8.16	$t_{(120)} = -6.73$	7.42×10^{-9} *	15.25	14.30: 16.21	23.41	21.21: 25.61
Aggression/Opposition	−4.42	$t_{(119)} = -4.11$	0.000109 *	5.15	4.30: 6.01	9.57	7.63: 11.51
Hyperactivity/Inattention	−7.7	$t_{(117)} = -5.51$	6.5×10^{-7} *	6.58	5.47: 7.69	14.28	11.75: 16.80
Depression/Anxiety	−1.34	$t_{(95.8)} = -1.31$	0.192	4.91	4.09: 5.73	6.25	4.41: 8.08
Physical Health	−0.798	$t_{(102)} = -1.35$	0.192	2.34	1.87: 2.81	3.14	2.07: 4.21

Note: BRIEF and PBS raw scores are reported. Diff Means indicates the difference in group mean estimates. Marg Mean indicates the estimated marginal means for each group. Abbreviations: GNE, Gene-Non-Expanded group; JOHD, Juvenile-Onset Huntington's Disease group. * indicates False Discovery Rate (FDR) *p*-adjusted < 0.0005.

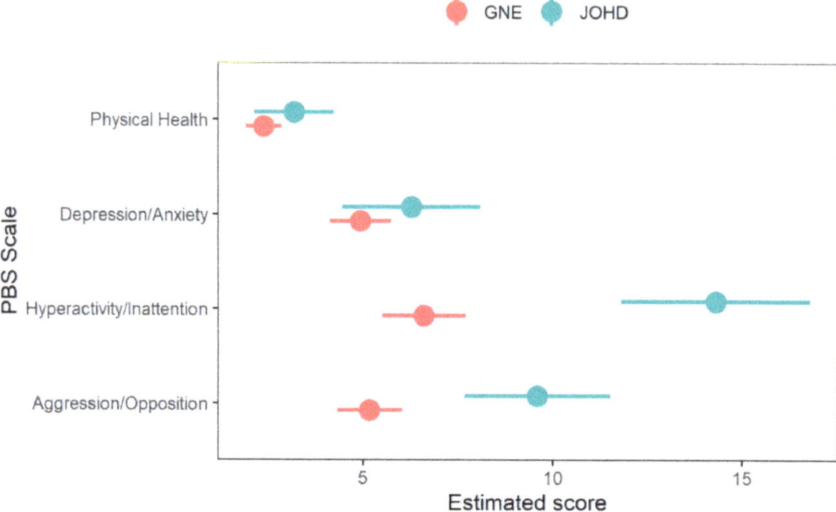

Figure 2. Differences in Pediatric Behavior Scale (PBS) subscale scores between JOHD (blue) and GNE (red) participants. The x-axis shows age- and sex-adjusted estimates from mixed linear effects models after controlling for random effects of repeated measures and family ID. The y-axis shows subscales of the Pediatric Behavior Scale-short form (PBS). The larger circles represent the means and horizontal lines indicate 95% confidence limits.

3.3. Genetic Expansion Correlations

Within JOHD, all measures had a negative correlation with CAG repeat length, however this reached significance for BRIEF Inhibit ($p = 0.048$), Plan/Organize ($p = 0.034$), and Initiate ($p = 0.013$) subscales of the BRIEF and the Aggression/Opposition ($p = 0.038$) scale of the PBS (see Table 3). A negative association indicates that JOHD participants with the highest CAG repeat tended to have the lowest behavioral scores. All other BRIEF and PBS subscales were not significantly predicted by CAG repeat length.

Table 3. Genetic expansion effects in JOHD.

Variable	Coefficient	95% CI	t-Value (df)	p-Value
Emotional Control	−0.08	−0.35: 0.19	$t_{(47)} = -0.615$	0.548
Inhibit	−0.23	−0.45: −0.01	$t_{(47)} = -2.08$	0.0479 *
Shift	−0.04	−0.18: 0.11	$t_{(47)} = -0.493$	0.627
Working Memory	−0.24	−0.53: 0.04	$t_{(47)} = -1.71$	0.102
Plan/Organize	−0.26	−0.50: −0.03	$t_{(47)} = -2.24$	0.0343 *
Initiate	−0.18	−0.32: −0.05	$t_{(46)} = -2.68$	0.0126 *
Self-Monitor	−0.16	−0.33: 0.01	$t_{(46)} = -1.87$	0.0723
Organization of Materials	−0.15	−0.36: 0.06	$t_{(46)} = -1.43$	0.166
Aggression/Opposition	−0.30	−0.57: −0.03	$t_{(47)} = -2.21$	0.0379 *
Hyperactivity/Inattention	−0.23	−0.51: 0.05	$t_{(47)} = -1.68$	0.114

Note: Statistics based on measures that were significantly different between GNE and JOHD groups. JOHD, Juvenile-Onset Huntington's Disease group. Higher means indicate more behavioral/executive problems on the BRIEF and PBS. * indicates *p*-value < 0.05.

3.4. BRIEF-A Report Type Differences

In total, there were 35 observations for BRIEF-A (JOHD = 13, GNE = 22; see Supplementary Table S3). The difference score in the GNE group was generally close to 0, except for Inhibit (p = 0.003) and Working Memory (p = 0.002), where informants reported more problems than participants. In contrast, in the JOHD group the difference scores were consistently different from 0, with informants reporting more problems than JOHD patients (all FDR < 0.05; see Figure 3 and Table 4). Age and sex did not have significant effects on the difference between informant- and self-reported scores.

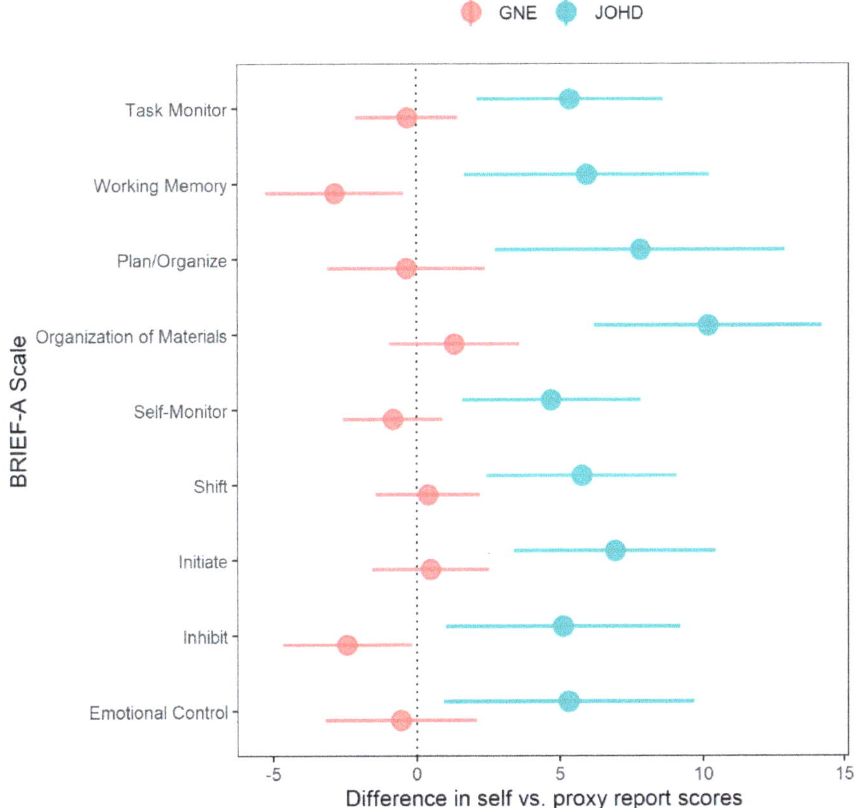

Figure 3. Differences in discrepancies between informant- and self-reported BRIEF-Adult (BRIEF-A) subscale scores between JOHD (blue) and GNE (red) participants. The x-axis shows the difference between informant and self-reported scores (calculated as informant—self) on the Behavioral Rating Inventory of Executive Function-Adult form (BRIEF-A); age- and sex-adjusted estimates are plotted from mixed linear effects models after controlling for random effects of family ID. The larger circles represent the means and horizontal lines indicate 95% confidence limits. The vertical black line at 0 marks no difference between informant and self-report scores.

Table 4. BRIEF-A model statistics and marginal means.

Variable	Diff Means	*t*-Value (df)	FDR	Marg Mean GNE	95% CI GNE	Marg Mean JOHD	95% CI JOHD
Emotional Control	−5.88	$t_{(18.6)} = -2.13$	0.0466 *	−0.58	3.25: 2.09	5.30	0.73: 9.87
Inhibit	−7.55	$t_{(25.1)} = -3.06$	0.0117 *	−2.45	−4.72: −0.18	5.10	0.96: 9.24
Initiate	−6.45	$t_{(20.4)} = -2.96$	0.0117 *	0.47	−1.60: 2.54	6.92	3.29: 10.54
Shift	−5.39	$t_{(25.2)} = -2.68$	0.0145 *	0.37	−1.48: 2.22	5.76	2.38: 9.13
Self-Monitor	−5.56	$t_{(24.6)} = -2.93$	0.0117 *	−0.85	−2.60: 0.90	4.71	1.55: 7.88
Organization of Materials	−8.88	$t_{(20.7)} = -3.6$	0.0113 *	1.29	−1.02: 3.61	10.18	6.07: 14.29
Plan/ Organize	−8.18	$t_{(25.5)} = -2.71$	0.0145 *	−0.38	−3.17: 2.41	7.80	2.67: 12.93
Working Memory	−8.81	$t_{(24.4)} = -3.37$	0.0113 *	−2.87	−5.30: −0.45	5.94	1.59: 10.28
Task Monitor	−5.71	$t_{(24.9)} = -2.89$	0.0117 *	−0.35	−2.16: 1.47	5.37	2.06: 8.67

Note: Statistics based on group differences between the difference in BRIEF-Adult report type, as calculated by informant-report scores—self-report scores. Diff Means indicates the difference in group mean estimates. Marg Mean indicates the estimated marginal means for each group. GNE, gene-non-expanded group; JOHD, juvenile-onset Huntington's Disease group. Higher means indicate more behavioral/executive problems on the BRIEF-A. * indicates FDR *p*-adjusted < 0.05.

4. Discussion

The Kids-JOHD study is the first ever prospective, longitudinal study of this ultra-rare population. Therefore, this is the first analysis of behavioral symptoms of JOHD that measures behavior on a continuum, rather than using reports of behavioral issues from retrospective analyses of medical records or qualitatively by parents and caretakers of these patients [3,5,8,9,11,16]. Our findings provide quantitative support for the notion that behavioral dysfunction is prevalent among persons with JOHD [3,5,8,9,11,16]. Specific commonly reported symptoms in JOHD include aggressive and oppositional behavior, and difficulties with attention; consistent with this, the largest group differences we found were in parental reports of Hyperactivity/Inattention followed by Aggression/Opposition [3,5,8,10,16].

Importantly, JOHD patients did not exhibit significant anxiety and depression. Different from externalizing behaviors that are easy for others to see, such as aggression and impulsivity, internalizing behaviors associated with subjective feelings of mood and being nervous may be harder to rate objectively [17]. Regardless of the inherent issues in parent reports of internalizing symptoms, it is clear that these symptoms were no more frequent in the JOHD subjects compared to the GNE children, supporting the notion that internalizing behaviors are not significantly affected in JOHD.

In the present study, mutant *HTT* CAG repeat expansion length was negatively correlated with all measures, but reached statistical significance with Inhibit, Plan/Organize, Initiate, and Aggression/Opposition indicating that patients with longer repeat lengths (typically resulting in childhood-onset JOHD [18]) exhibit fewer problems in these domains, while patients with short repeats (typically resulting in adolescent-onset JOHD [18]) exhibit more problem behaviors. These findings align with current reports that older-onset JOHD patients exhibit more behavioral issues than younger-onset JOHD patients [7]. In addition, opposition and aggression behavior normally peak in adolescence; therefore, an active brain disease during a time in which these behaviors normally peak may be one potential rationale for the adolescent onset having greater behavioral disturbance. However, in review of the reports from caregivers of those with childhood onset, problems with aggression and opposition were common early in the course of the disease, years prior to diagnosis. It may be that by the time of motor onset, the childhood onset patients have moved past a period of externalizing behavior [1,7].

Analyses evaluating differences between informant and self-reported measures of behavioral regulation and executive function indicated a possible lack of insight among JOHD patients in their behaviors. While the GNE group had similar scores between informant and self-reports, the JOHD participants consistently rated themselves as having fewer behavioral and executive functioning problems than what was reported by their informants. Limited insight into behavioral and cognitive difficulties is a known feature in patients with AOHD [19]. These results suggest that informant ratings are crucial when quantifying behavioral and executive dysfunction in JOHD.

Parents often report that cognitive and behavior issues are the first harbinger of change and can occur sometimes years prior to final motor diagnosis. Some families in the JOHD community have lobbied for using behavioral changes as a diagnostic criteria for disease [8]. This would be inappropriate for several reasons. Any symptom utilized for diagnosis has to be sensitive and specific to the disease. Although all of the subjects here are already motor manifest, it is important to point out that behavioral symptoms are not present in all patients, therefore these behavioral ratings are not sensitive to the presence of JOHD. Secondly elevated behavioral ratings are not specific to JOHD [6]. There were many children in the GNE group with elevated scores (in fact the two highest scores in the entire sample on hyperactivity and inattention from the PBS came from GNE participants). Although as a group, the JOHD sample had elevated scores compared to GNE, the presence of elevated scores in any one individual is not specific to JOHD. This underscores the notion that changes in behavior should not be utilized for diagnostic purposes for children at risk for AOHD or JOHD.

This study was not without limitations. First, with JOHD being an extremely rare disorder affecting only 1–10% of individuals with HD [1], our sample was limited to 21 individuals with JOHD; however, we leveraged an accelerated longitudinal design to increase the number of observations. Second, parent and informant ratings are objective measures of behavior. Since parents of JOHD are aware of their diagnosis, they may be biased in reporting their child as having greater symptoms, simply knowing that it is commonly known amongst these families that behavioral disturbances occur in children with JOHD. Finally, since self-reported measures were not used in this study for individuals younger than 18, we relied on a small sample for our self vs. informant analyses.

5. Conclusions

In this study, patients with juvenile-onset Huntington's Disease (JOHD) exhibited significant behavioral problems relative to gene non-expanded (GNE) counterparts. Those participants with longer CAG repeats (earlier age of onset) had fewer behavioral problems compared to those with relatively shorter repeats (later age of onset). Lack of insight may have prohibited adult patients with JOHD from providing reliable assessments of their behavioral problems. Further research should be conducted with larger samples to create a JOHD-specific behavioral rating measure including both self and proxy measures that may be used as markers for clinical trials and treatments.

Supplementary Materials: The following are available online at http://www.mdpi.com/2076-3425/10/8/543/s1, Table S1: BRIEF subscale statistics, Table S2: PBS subscale statistics, Table S3: BRIEF-Adult subscale statistics.

Author Contributions: Conceptualization, K.E.L., E.v.d.P. and P.N.; methodology, K.E.L., E.v.d.P., and P.N.; software, K.E.L. and E.v.d.P.; validation, K.E.L., E.v.d.P., and P.N.; formal analysis, K.E.L. and E.v.d.P.; investigation, A.M.C., E.E., E.M. and P.E.-P.; resources, P.N.; data curation, A.M.C., E.E., E.M., P.E.-P.; writing—original draft preparation, K.E.L., A.M.C., E.v.d.P.; writing—review and editing, A.L.C., E.E., E.M., P.E.-P. and P.N.; visualization, K.E.L. and E.v.d.P.; supervision, E.v.d.P. and P.N.; project administration, P.N.; funding acquisition, P.N. All authors have read and agreed to the published version of the manuscript.

Funding: This study was supported by National Institute of Neurological Disorders and Stroke (NINDS; R01NS055903). The APC was funded by National Institute of Neurological Disorders and Stroke (NINDS; R01NS055903).

Acknowledgments: In this section you can acknowledge any support given which is not covered by the author contribution or funding sections. This may include administrative and technical support, or donations in kind (e.g., materials used for experiments).

Conflicts of Interest: The authors declare no conflict of interest. The funders had no role in the design of the study; in the collection, analyses, or interpretation of data; in the writing of the manuscript, or in the decision to publish the results.

References

1. Barker, R.A.; Squitieri, F. The clinical phenotype of juvenile Huntington's Disease. In *Juvenile Huntington's Disease and Other Trinucleotide Repeat Disorders*; Quarrell, O.W.J., Brewer, H.M., Squitieri, F., Barker, R.A., Nance, M.A., Landwehrmeyer, G.B., Eds.; Oxford University Press: New York, NY, USA, 2009; pp. 39–50.
2. Fusilli, C.; Migliore, S.; Mazza, T.; Consoli, F.; De Luca, A.; Barbagallo, G.; Ciammola, A.; Gatto, E.M.; Cesarini, M.; Etcheverry, J.L.; et al. Biological and clinical manifestations of juvenile Huntington's disease: A retrospective analysis. *Lancet Neurol.* **2018**, *17*, 986–993. [CrossRef]
3. Gonzalez-Alegre, P.; Afifi, A.K. Clinical characteristics of childhood-onset (juvenile) Huntington disease: Report of 12 patients and review of the literature. *J. Child. Neurol.* **2006**, *21*, 223–229. [CrossRef] [PubMed]
4. Moser, A.D.; Epping, E.; Espe-Pfeifer, P.; Martin, E.; Zhorne, L.; Mathews, K.; Nance, M.; Hudgell, D.; Quarrell, O.; Nopoulos, P. A survey-based study identifies common but unrecognized symptoms in a large series of juvenile Huntington's disease. *Neurodegener. Dis. Manag.* **2017**, *7*, 307–315. [CrossRef] [PubMed]
5. Ribai, P.; Nguyen, K.; Hahn-Barma, V.; Gourfinkel-An, I.; Vidailhet, M.; Legout, A.; Dode, C.; Brice, A.; Durr, A. Psychiatric and cognitive difficulties as indicators of juvenile Huntington disease onset in 29 patients. *Arch. Neurol.* **2007**, *64*, 813–819. [CrossRef] [PubMed]
6. Smith, J.A.; Brewer, H.M.; Eatough, V.; Stanley, C.A.; Glendinning, N.W.; Quarrell, O.W. The personal experience of juvenile Huntington's disease: An interpretative phenomenological analysis of parents' accounts of the primary features of a rare genetic condition. *Clin. Genet.* **2006**, *69*, 486–496. [CrossRef] [PubMed]
7. Cronin, T.; Rosser, A.; Massey, T. Clinical Presentation and Features of Juvenile-Onset Huntington's Disease: A Systematic Review. *J. Huntingt. Dis.* **2019**, *8*, 171–179. [CrossRef] [PubMed]
8. Nance, M.A. The treatment of juvenile Huntington's disease. In *Juvenile Huntington's Disease and Other Trinucleotide Repeat Disorders*; Quarrell, O.W.J., Brewer, H.M., Squitieri, F., Barker, R.A., Nance, M.A., Landwehrmeyer, G.B., Eds.; Oxford University Press: New York, NY, USA, 2009; pp. 151–166.
9. Poniatowska, R.; Habib, N.; Krawczyk, R.; Rakowicz, M.; Niedzielska, K.; Hoffman, D.; Jakubowska, T.; Bogusławska, R. Correlation between magnetic resonance and genetic, clinical, neurophysiological and neuropsychological studies of patients with juvenile form of Huntington's disease. *Case Rep. Clin. Pract. Rev.* **2001**, *2*, 140–146.
10. Quigley, J. Juvenile Huntington's Disease: Diagnostic and Treatment Considerations for the Psychiatrist. *Curr. Psychiatry Rep.* **2017**, *19*, 9. [CrossRef] [PubMed]
11. Rasmussen, A.; Macias, R.; Yescas, P.; Ochoa, A.; Davila, G.; Alonso, E. Huntington disease in children: Genotype-phenotype correlation. *Neuropediatrics* **2000**, *31*, 190–194. [CrossRef] [PubMed]
12. Van der Plas, E.; Langbehn, D.R.; Conrad, A.L.; Koscik, T.R.; Tereshchenko, A.; Epping, E.A.; Magnotta, V.A.; Nopoulos, P.C. Abnormal brain development in child and adolescent carriers of mutant huntingtin. *Neurology* **2019**, *93*, e1021–e1030. [CrossRef] [PubMed]
13. Vijayakumar, N.; Mills, K.L.; Alexander-Bloch, A.; Tamnes, C.K.; Whittle, S. Structural brain development: A review of methodological approaches and best practices. *Dev. Cogn. Neurosci.* **2018**, *33*, 129–148. [CrossRef] [PubMed]
14. Gioia, G.A.; Isquith, P.; Guy, S.; Kenworthy, L. *Behavior Rating Inventory of Executive Function: Professional Manual*; Psychological Assessment Resources, Inc.: Odessa, FL, USA, 2000.
15. Lindgren, S.; Koeppl, G. Assessing Child Behavior Problems in a Medical Setting: Development of the Pediatric Behavior Scale. In *Advances in Behavioral Assessment of Children and Families*; Prinz, E., Ed.; JAI: Greenwich, CT, USA, 1987; pp. 57–90.
16. Quarrell, O.W.; Nance, M.A.; Nopoulos, P.; Paulsen, J.S.; Smith, J.A.; Squitieri, F. Managing juvenile Huntington's disease. *Neurodegener. Dis. Manag.* **2013**, *3*, 267–276. [CrossRef] [PubMed]
17. Klonsky, E.D.; Oltmanns, T.F. Informant-reports of personality disorder: Relation to self-reports and future research directions. *Clin. Psychol. Sci. Pract.* **2002**, *9*, 300–311. [CrossRef]

18. Quarrell, O.; O'Donovan, K.L.; Bandmann, O.; Strong, M. The Prevalence of Juvenile Huntington's Disease: A Review of the Literature and Meta-Analysis. *PLoS Curr.* **2012**, *4*, e4f8606b8742ef8603. [CrossRef] [PubMed]

19. McCusker, E.; Loy, C.T. The many facets of unawareness in huntington disease. *Tremor Other Hyperkinet. Mov.* **2014**, *4*, 257. [CrossRef]

Communication

Subcortical T1-Rho MRI Abnormalities in Juvenile-Onset Huntington's Disease

Alexander V. Tereshchenko [1,†], Jordan L. Schultz [1,2,3,*,†], Ansley J. Kunnath [4], Joel E. Bruss [1,2], Eric A. Epping [1], Vincent A. Magnotta [1,5] and Peg C. Nopoulos [1,3,6]

[1] Department of Psychiatry, Carver College of Medicine at the University of Iowa, Iowa City, IA 52242, USA; alexander-tereshchenko@uiowa.edu (A.V.T.); joel-bruss@uiowa.edu (J.E.B.); eric-epping@uiowa.edu (E.A.E.); Vincent-magnotta@uiowa.edu (V.A.M.); peggy-nopoulos@uiowa.edu (P.C.N.)

[2] Department of Neurology, Carver College of Medicine at the University of Iowa, Iowa City, IA 52242, USA

[3] Department of Pharmacy Practice and Science, College of Pharmacy, University of Iowa, Iowa City, IA 52242, USA

[4] Department of Cell Biology and Neuroscience, Rutgers University, New Brunswick, NJ 08854, USA; Ansley.j.kunnath@Vanderbilty.edu

[5] Department of Radiology, Carver College of Medicine at the University of Iowa, Iowa City, IA 52242, USA

[6] Department of Pediatrics, Stead Family Children's Hospital at the University of Iowa, Iowa City, IA 52242, USA

* Correspondence: jordan-schultz@uiowa.edu; Tel.: +1-319-384-9388

† These authors contributed equally to this work.

Received: 15 July 2020; Accepted: 6 August 2020; Published: 8 August 2020

Abstract: Huntington's disease (HD) is a fatal neurodegenerative disease caused by the expansion of cytosine-adenine-guanine (CAG) repeats in the *huntingtin* gene. An increased CAG repeat length is associated with an earlier disease onset. About 5% of HD cases occur under the age of 21 years, which are classified as juvenile-onset Huntington's disease (JOHD). Our study aims to measure subcortical metabolic abnormalities in JOHD participants. T1-Rho ($T_{1\rho}$) MRI was used to compare brain regions of 13 JOHD participants and 39 controls. Region-of-interest analyses were used to assess differences in quantitative $T_{1\rho}$ relaxation times. We found that the mean relaxation times in the caudate ($p < 0.001$), putamen ($p < 0.001$), globus pallidus ($p < 0.001$), and thalamus ($p < 0.001$) were increased in JOHD participants compared to controls. Furthermore, increased $T_{1\rho}$ relaxation times in these areas were significantly associated with lower volumes amongst participants in the JOHD group. These findings suggest metabolic abnormalities in brain regions previously shown to degenerate in JOHD. We also analyzed the relationships between mean regional $T_{1\rho}$ relaxation times and Universal Huntington's Disease Rating Scale (UHDRS) scores. UHDRS was used to evaluate participants' motor function, cognitive function, behavior, and functional capacity. Mean $T_{1\rho}$ relaxation times in the caudate ($p = 0.003$), putamen ($p = 0.005$), globus pallidus ($p = 0.009$), and thalamus ($p = 0.015$) were directly proportional to the UHDRS score. This suggests that the $T_{1\rho}$ relaxation time may also predict HD-related motor deficits. Our findings suggest that subcortical metabolic abnormalities drive the unique hypokinetic symptoms in JOHD.

Keywords: juvenile-onset Huntington's disease; T1-Rho; neuroimaging

---

1. Introduction

Huntington's disease (HD) is caused by the expansion of cytosine-adenine-guanine (CAG) repeats in the *huntingtin* gene, and the CAG repeat length is associated with an earlier disease onset [1]. About 5% of HD cases occur before the age of 21 years, which are classified as juvenile-onset Huntington's disease (JOHD) [1]. Adult-onset HD (AOHD) is characterized by involuntary movements (chorea), whereas JOHD often causes bradykinesia [1].

Patients with JOHD experience similar patterns of neurodegeneration as patients with AOHD with some distinct exceptions. Specifically, patients with JOHD have proportionally larger cerebellar volumes relative to controls, and they do not experience thinning of the motor cortex [2] as is seen in AOHD [3]. Since these brain regions are relatively spared, it has been hypothesized that they may play a compensatory role in JOHD. Additionally, both the cerebellum and motor strip play key roles in maintaining motor control, and patients with JOHD experience markedly different motor symptoms compared to patients with AOHD. However, these hypotheses are based solely on volumetric data, and little is known regarding the functional activity of the neurons in these brain regions. T_1 relaxation in the rotating frame ($T_{1\rho}$) is a novel neuroimaging modality that allows for analysis of biochemical changes in the brain that are undetectable by existing MRI techniques. The use of a spin-locking radio-frequency field increases sensitivity to proton exchange, which is influenced by pH, glucose, glutamate, water, and proteins. $T_{1\rho}$ MRI has previously been used to characterize progressive changes in other neurodegenerative diseases. Specifically, cross-sectional studies have shown that patients with Alzheimer's disease have increased $T_{1\rho}$ relaxation times in the hippocampus compared to controls [4–7]. Similar findings of longer $T_{1\rho}$ relaxation times have been reported in patients with Parkinson's disease [4,8,9], bipolar disorder [10], and multiple sclerosis [11]. These studies all employed cross-sectional designs. $T_{1\rho}$ relaxation times were also shown to increase in the striatum of participants with premanifest patients with AOHD [12].

To the best of our knowledge, $T_{1\rho}$ relaxation times have never been evaluated in patients with JOHD, but may be an important biomarker for progression in this unique subset of patients with HD. Longitudinal changes in the volume of the caudate and putamen have been studied extensively in AOHD and may serve as a valid biomarker for disease progression in that patient population. However, striatal degeneration likely begins very early in life in JOHD and may not change significantly enough over time to serve as a valid biomarker of disease progression for future clinical trials. Here, we aimed to determine if $T_{1\rho}$ relaxation times in subcortical regions of the brain in patients with JOHD were significantly different from healthy controls.

2. Materials and Methods

2.1. Participants

Participants included in these analyses were enrolled in the Kids-HD and Kids-JOHD studies [2,13–15]. These were longitudinal neuroimaging studies that ran in parallel to one another. The Kids-HD study recruited participants between the ages of 6–26 who were at risk for inheriting the gene that causes HD based on their family history (i.e., a parent or grandparent with confirmed HD). The Kids-HD study also recruited healthy control participants. All participants in the Kids-HD study underwent genetic testing for research purposes only. For the present analyses, the control group consisted of participants from the Kids-HD study who had molecular confirmation of a CAG repeat length < 36, ensuring that our control group did not include pre-symptomatic patients with AOHD. The Kids-JOHD study recruited participants who had been deemed to have Juvenile-Onset HD (JOHD) by their neurologist and had molecular confirmation of having a CAG repeat length of 36 or above. For these analyses, a participant was considered to have JOHD if they had a total motor score from the Unified Huntington's Disease Rating Scale (UHDRS) [16] of ≥20 prior to the age of 21. The UHDRS is sensitive to developmental motor changes such that younger children will show higher scores than older children. Therefore, even in a large cohort of children at risk, but who did not inherit the gene expansion, the UHDRS can be as high as 15 [15]. Therefore, a cutoff of 20 on the UHDRS was used to ensure that all participants had confirmed motor-manifest JOHD.

Given the longitudinal nature of these studies, some participants had more than one neuroimaging study conducted. Specifically, there were 11 participants with JOHD that made up 13 visits. Our control group in these analyses consisted of 38 participants that had the necessary neuroimaging done, which included both anatomic (T_1- and T_2-weighted) and metabolic ($T_{1\rho}$) images. This consisted of participants who were at risk for inheriting the gene mutation that causes HD but who were found to not carry this gene, as well as healthy control participants without a family history of HD. There were 39 neuroimaging studies amongst this group of 38 participants.

Clinical measures were collected on all participants. As noted previously, all participants received a total motor score as measured using the UHDRS [16]. Motor symptoms were also quantified in the JOHD group using the modified Juvenile HD Motor Rating Scale (JOHDRS) [17]. This rating scale provides additional evaluation of the unique hypokinetic symptoms associated with JOHD. We also calculated a disease burden score (Age × CAG—35.5)) [18] and disease duration (age at time of assessment—age at time of JOHD clinical diagnosis) for JOHD participants.

Signed informed consent was obtained before beginning the study, per the Institutional Review Board at the University of Iowa. All experiments were performed following the guidelines outlined in the Belmont Report. Genetic testing was done for research purposes only. The results were made available to one research team member, and all other team members were blind to these genetic results. The genetic testing results were not revealed to participants or their families.

2.2. Data Acquisition

High-resolution magnetic resonance images were collected on a 3T Siemens TIM Trio scanner (Siemens Medical Solutions, Erlangen, Germany) using a 12-channel receiver head coil. Whole-brain T_1- and T_2-weighted anatomical acquisitions with a 1.0 mm isotropic spatial resolution were acquired first. T_1-weighted images were collected using a 3D magnetization-prepared rapid gradient echo sequence with the following parameters: Coronal orientation; field-of-view = 25.6 × 25.6 × 25.6 cm; sampling matrix = 256 × 256 × 256; repetition time (TR)TR/echo time (TE)/inversion time (TI) = 2530/2.8/909 ms; flip angle = 10°; bandwidth = 180 Hz/pixel; and Acceleration Factor (R) = 2 GeneRalized Autocalibrating Partial Parallel Acquisition (GRAPPA). T_2-weighted images were collected using a 3D variable flip angle spin-echo sequence with the following parameters: TE = 430 ms; TR = 4800 ms; number of echos = 137; bandwidth = 592 Hz/pixel; matrix = 256 × 256 × 170; field-of-view (FOV) = 25.6 × 25.6 × 22 cm; and R = 2 GRAPPA. Next, quantitative parametric imaging was conducted to acquire $T_{1\rho}$ relaxation times. $T_{1\rho}$ mapping was performed using a coronal segmented three-dimensional (3D) gradient echo sequence with spin-lock pulses (TE = 2.5 ms; TR = 5.6 ms, FOV = 220 × 220 × 200 mm; sampling matrix = 128 × 128 × 40; fractional anisotropy (FA) = 10 degrees; integrated parallel acquisition techniques (IPAT) = 2; spin-lock frequency = 330 Hz; spin-lock times = 10 and 55 ms).

2.3. Image Analysis

The BRAINS AutoWorkup was used to perform the anatomical image analysis by combining information available from the T_1- and T_2-weighted images as described in Pierson et al. [19]. Briefly, the BRAINS AutoWorkup includes the following steps: (1) AC-PC alignment; (2) bias field correction; (3) tissue classification; and (4) anatomical labeling. The anatomical regions-of-interest used for this study include the caudate, putamen, globus pallidus, thalamus, hippocampus, and anterior cerebellum. These regions-of-interest are defined automatically using a neural network-based segmentation [20]. These regions have been shown to have a degree of reliability with a manual rater (Jaccard index ~0.80).

To estimate the $T_{1\rho}$ map, the individual spin-lock images were co-registered using a rigid registration with the Advanced Normalization Tools (ANTs) software [21] to account for subject motion between the two acquisitions. The resulting aligned spin-lock images were then used to calculate a $T_{1\rho}$ map by fitting the 10 and 55 ms spin-lock time (TSL) image signals (S_0 and S_{TSL}) to the following mono-exponential decay model:

$$S_{TSL} = S_0 \left(e^{-TSL/T_{1\rho}} \right) \tag{1}$$

The decay model was fit using the "MR Parameter Map Suite" implemented using Insight Segmentation and Registration Toolkit (ITK) [22] and available from the InsightJournal [23]. The resulting rigid body transform was then used to resample the $T_{1\rho}$ map to a 1 mm isotropic resolution using linear interpolation. The $T_{1\rho}$ maps were then thresholded at 400 ms to remove the contribution of cerebrospinal fluid. The defined regions-of-interest generated from the BRAINS AutoWorkup were then used to estimate the mean $T_{1\rho}$ relaxation times for each region from non-zero voxels in the thresholded $T_{1\rho}$ relaxation time maps.

2.4. Statistical Analysis

The primary outcomes were mean differences in $T_{1\rho}$ relaxation times from the defined regions-of-interest between groups. We used linear mixed effects models to investigate estimated mean differences in $T_{1\rho}$ relaxation times of the six brain regions above between groups. Our models were controlled for age and sex and included a random effect per participant.

For the secondary analyses, we identified any of the regions-of-interest that demonstrated significant group differences in $T_{1\rho}$ relaxation times from the primary analysis. Amongst those regions-of-interest, we used linear mixed effects regression analyses to investigate the relationship between CAG repeat length and regional brain volumes and $T_{1\rho}$ relaxation times. These models were controlled for age and sex and included a random effect per participant. The models investigating the relationship between brain volume and $T_{1\rho}$ relaxation times were also controlled for intracranial volume (ICV). All of these analyses were performed amongst the JOHD participants only. We then performed a similar analysis to assess the relationship between $T_{1\rho}$ relaxation times and disease burden scores while controlling for sex and a participant random effect. Age was not included as age is used to calculate the disease burden score [18]. Next, we assessed whether $T_{1\rho}$ relaxation times predicted the total motor score, as assessed by the UHDRS and the JOHDR [17], while controlling for age and CAG repeat length. Lastly, we investigated the relationship between the calculated disease duration and $T_{1\rho}$ relaxation times. Again, we performed linear mixed effects regression analyses that controlled for age and sex and a random effect per participant. RStudio (Version 3.6.2, RStudio, PBC, Boston, MA, USA) was used for all statistical analyses and a *p*-value of <0.05 was considered significant for all analyses.

3. Results

3.1. Primary Outcomes

Participant demographics are tabulated in Table 1.

Table 1. Demographics by groups.

	JOHD Group	**Controls**	***p*-Value**
N (Visits)	11 (13)	38 (39)	NA
Male, % (n)	36.4 (4)	39.5 (15)	1
Age (yrs), Mean ± SD	15.39 ± 5.1	15.56 ± 3.82	0.905
CAG Repeats, Mean ± SD	72.82 ± 10.31	19.5 ± 3.85	<0.001
Disease Burden Score, Mean ± SD	540.99 ± 148.11	NA	NA
Disease Duration (yrs), Mean ± SD	3.1 ± 2.61	NA	NA
UHDRS, Mean ± SD	55.27 ± 21.54	NA	NA
JOHDRS, Mean ± SD	14.91 ± 6.71	NA	NA

For the primary analysis, the JOHD group had significantly longer $T_{1\rho}$ relaxation times in the caudate, putamen, globus pallidus, and thalamus compared to the control group (Table 2), indicating significant neuronal damage in these areas in patients with JOHD. However, there were no significant group differences regarding $T_{1\rho}$ relaxation times in the hippocampus and cerebellum, indicating no difference in neuronal damage in these areas.

Table 2. Regional $T_{1\rho}$ relaxation times—juvenile-onset Huntington's disease (JOHD) participants.

Region	Control $T_{1\rho}$ (ms)	JOHD $T_{1\rho}$ (ms)	Beta-Coefficient	*p*-Value
Caudate, mean ± SD	78.34 ± 6.35	106.42 ± 14.21	27.9	<0.001
Putamen, mean ± SD	71.25 ± 2.56	81.79 ± 3.53	10.37	<0.001
Globus pallidus, mean ± SD	64.13 ± 2.79	68.52 ± 4.05	4.62	<0.001
Thalamus, mean ± SD	73.86 ± 2.46	76.38 ± 2.37	2.53	<0.001
Hippocampus, mean ± SD	86.12 ± 8.5	87.65 ± 3.49	1.76	0.484
Anterior Cerebellum mean ± SD	92.12 ± 18.81	94.13 ± 11.55	1.94	0.732

3.2. Secondary Outcomes

Higher $T_{1\rho}$ relaxation times were associated with lower volumes in the caudate (t = −2.46, p = 0.039), putamen (t = −5.63, p = 0.0006), and globus pallidus (t = −2.32, p = 0.0491). There was a negative relationship between $T_{1\rho}$ relaxation times and volume of the thalamus, but the results did not reach statistical significance (t = −1.92, p = 0.0912). Next, we demonstrated significant positive relationships between CAG repeat length and $T_{1\rho}$ relaxation times in the caudate (t = 3.02, p = 0.018), putamen (t = 3.73, p = 0.006), globus pallidus (t = 7.88, p < 0.0001), and thalamus (t = 2.68, p = 0.026) (Figure 1A–D). The positive relationship indicates that the higher CAG repeat length is associated with increased neuronal damage in these brain regions.

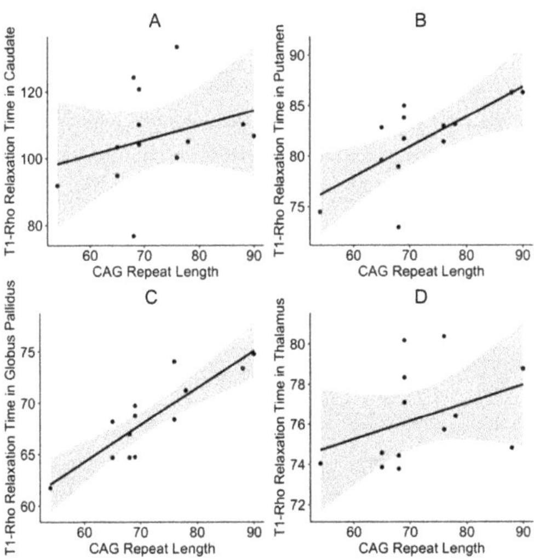

Figure 1. CAG repeat length significantly predicts $T_{1\rho}$ relaxation times in the (**A**) caudate, (**B**) putamen, (**C**) globus pallidus, and (**D**) thalamus. Results show raw data points, the fitted regression line of the model, and 95% confidence interval. CAG: Cytosine-adenine-guanine.

While the relationship between CAG repeat length and $T_{1\rho}$ relaxation times in the caudate and thalamus were statistically significant, there seemed to be some outlying data that could have influenced the results. Given the small sample size of patients, we performed unplanned follow-up analyses to account for the potential influence of outliers on the results. Specifically, we repeated the analyses using robust linear mixed effects regression using the "robust" package in R. The relationship between CAG repeat length and $T_{1\rho}$ relaxation times in the caudate (t = 2.91, p = 0.021) and thalamus (t = 3.18, p = 0.012) remained significant after accounting for potential outliers.

Next, we assessed the relationship between $T_{1\rho}$ relaxation times and disease burden scores. The disease burden scores significantly predicted $T_{1\rho}$ relaxation times in the caudate (t = 3.97, p = 0.003) and thalamus (t = 3.07, p = 0.012), but not in the globus pallidus (t = 1.97, p = 0.081) or putamen (t = 1.95, *p* = 0.08). We also investigated the relationship between regional $T_{1\rho}$ relaxation times and motor function. We found that higher mean $T_{1\rho}$ relaxation times in the caudate (t = 3.55, *p* = 0.006), putamen (t = 3.26, *p* = 0.011), globus pallidus (t = 3.4, *p* = 0.008), and thalamus (t = 3.29, *p* = 0.042) were positively related to increased UHDRS scores (Figure 2A–D).

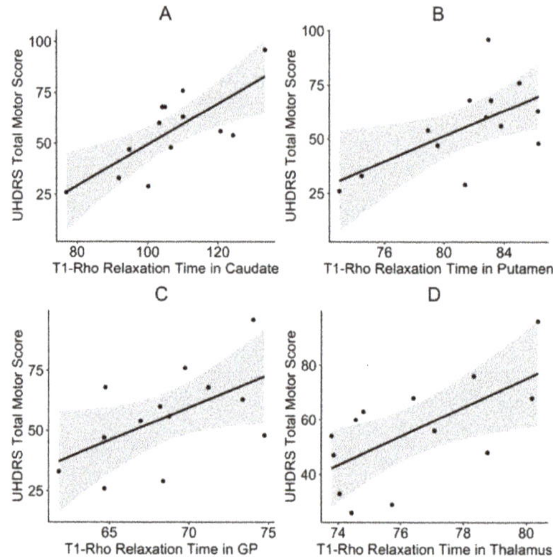

Figure 2. $T_{1\rho}$ relaxation times in the (**A**) caudate, (**B**) putamen, (**C**) globus pallidus, and (**D**) thalamus. Significantly predicted total motor scores as measured by the UHDRS. Results show raw data points, the fitted regression line of the model, and 95% confidence interval. GP: Globus pallidus; UHDRS: Unified Huntington's Disease Rating Scale.

Mean $T_{1\rho}$ relaxation times in the caudate (t = 3.08, *p* = 0.013) and putamen (t = 3.51, *p* = 0.007) were also directly proportional to the JOHDRS score, but not in the globus pallidus (t = 1.79, *p* = 0.131) or thalamus (t = 1.09, *p* = 0.357).

Lastly, we identified significant positive relationships between disease duration and $T_{1\rho}$ relaxation times in the caudate (t = 3.58, *p* = 0.008), putamen (t = 3.83, *p* = 0.01), and globus pallidus (t = 3.49, *p* = 0.007), but not in the thalamus (t = 1.09, *p* = 0.304). This suggests that $T_{1\rho}$ relaxation times may be able to track the disease course over time.

4. Discussion

In this study, we have utilized a novel neuroimaging method to demonstrate that metabolic abnormalities likely affect the caudate, putamen, globus pallidus, and thalamus in patients with JOHD. However, there were no significant group differences for the mean $T_{1\rho}$ relaxation time in the anterior cerebellum, an area which also controls motor function. Similarly, there were no significant differences in $T_{1\rho}$ relaxation times in the hippocampus between the JOHD and control groups. These results are interesting because the cerebellum is thought to be spared in HD and has even been found to be proportionally enlarged in JOHD [2]. The cerebellum has been hypothesized to play a compensatory role in HD and potentially in JOHD and these results may provide additional support for this theory [24].

The increases in subcortical $T_{1\rho}$ relaxation times were directly related to CAG repeat length in the JOHD group, increasing the likelihood that these findings are related to pathological changes. The disease burden score significantly predicted $T_{1\rho}$ relaxation times in the caudate and thalamus, but did not reach the level of significance for predicting relaxation times in the putamen and globus pallidus, despite a trend in that direction. Again, these findings indicate that $T_{1\rho}$ relaxation times seem to be indicative of disease severity and may be used as a unique measure of disease progression in JOHD. This is further supported by the finding that the longer duration of disease of JOHD was associated with significantly higher $T_{1\rho}$ relaxation times in the caudate, putamen, and globus pallidus.

Patients with JOHD often experience unique motor symptoms. Specifically, patients with JOHD may have less chorea, but more hypokinetic symptoms, including bradykinesia and dystonia [25]. These symptoms can be very difficult to treat and the underlying pathology of this difference between JOHD and AOHD is poorly understood. All of the subcortical regions that we analyzed were significantly and positively associated with the total motor score on the UHDRS. Additionally, relaxation times in the caudate and putamen significantly predicted total scores on the JOHD-specific motor assessment scale. This suggests that metabolic abnormalities in the striatum may drive the unique hypokinetic motor deficits of JOHD.

There are some potential limitations to this study. $T_{1\rho}$ MRI abnormalities in JOHD are not indicative of a specific metabolic dysfunction since it is sensitive to changes in several factors including pH, glucose, glutamate, water, and proteins. Further research using magnetic resonance spectroscopy may investigate specific molecular imbalances in the brain regions identified within this study. Additionally, $T_{1\rho}$ MRI images were captured last in a series of different neuroimaging tests, so JOHD participants with more severe symptoms were not able to complete $T_{1\rho}$ imaging. However, participants with very severe symptoms may have significantly diminished the striatum that is difficult to measure. Finally, the current study was limited by its small sample size due to the rarity of JOHD. As a result, further studies are required to confirm these results in an expanded patient population.

5. Conclusions

In conclusion, $T_{1\rho}$ MRI may be a valuable biomarker for monitoring disease progression and evaluating future clinical trials in JOHD. This novel imaging technique allows for high-resolution, quantitative analysis of subcortical metabolic changes. $T_{1\rho}$ MRI abnormalities in JOHD participants were found within the caudate, putamen, globus pallidus, and thalamus, and mean $T_{1\rho}$ relaxation times within these regions were predictive of disease severity and motor deficits.

Author Contributions: Conceptualization, A.V.T., J.L.S., A.J.K., J.E.B., V.A.M. and P.C.N.; methodology, A.V.T., J.L.S., A.J.K., J.E.B. and V.A.M.; software, A.V.T. and J.L.S.; validation, A.V.T., J.L.S. and P.C.N.; formal analysis, A.V.T., J.L.S. and P.C.N.; investigation, E.A.E. and P.C.N.; resources, V.A.M. and P.C.N.; data curation, E.A.E. and P.C.N.; writing—original draft preparation, A.V.T., J.L.S. and A.J.K.; writing—review and editing, J.E.B., E.A.E., V.A.M. and P.C.N.; visualization, J.L.S.; supervision, P.C.N.; project administration, P.C.N.; funding acquisition, P.C.N. All authors have read and agreed to the published version of the manuscript.

Funding: This study was supported by the National Institute of Neurological Disorders and Stroke (NINDS; R01NS055903). The APC was funded by the National Institute of Neurological Disorders and Stroke (NINDS; R01NS055903).

Conflicts of Interest: The authors declare no conflict of interest. The funders had no role in the design of the study; in the collection, analyses, or interpretation of data; in the writing of the manuscript, or in the decision to publish the results.

References

1. Nopoulos, P.C. Huntington disease: A single-gene degenerative disorder of the striatum. *Dialogues Clin. Neurosci.* **2016**, *18*, 91–98. [PubMed]
2. Tereshchenko, A.; Magnotta, V.; Epping, E.; Mathews, K.; Espe-Pfeifer, P.; Martin, E.; Dawson, J.; Duan, W.; Nopoulos, P. Brain structure in juvenile-onset Huntington disease. *Neurology* **2019**, *92*, e1939–e1947. [CrossRef] [PubMed]
3. Nopoulos, P.C.; Aylward, E.H.; Ross, C.A.; Johnson, H.J.; Magnotta, V.A.; Juhl, A.R.; Pierson, R.K.; Mills, J.; Langbehn, D.R.; Paulsen, J.S.; et al. Cerebral cortex structure in prodromal Huntington disease. *Neurobiol. Dis.* **2010**, *40*, 544–554. [CrossRef] [PubMed]
4. Haris, M.; Singh, A.; Cai, K.; Davatzikos, C.; Trojanowski, J.Q.; Melhem, E.R.; Clark, C.M.; Borthakur, A. T1rho (T1ρ) MR imaging in Alzheimer's disease and Parkinson's disease with and without dementia. *J. Neurol.* **2011**, *258*, 380–385. [CrossRef]

5. Haris, M.; Yadav, S.K.; Rizwan, A.; Singh, A.; Cai, K.; Kaura, D.; Wang, E.; Davatzikos, C.; Trojanowski, J.Q.; Melhem, E.R.; et al. T1rho MRI and CSF biomarkers in diagnosis of Alzheimer's disease. *Neuroimage Clin.* **2015**, *7*, 598–604. [CrossRef]

6. Haris, M.; McArdle, E.; Fenty, M.; Singh, A.; Davatzikos, C.; Trojanowski, J.Q.; Melhem, E.R.; Clark, C.M.; Borthakur, A. Early marker for Alzheimer's disease: Hippocampus T1rho ($T_{1\rho}$) estimation. *J. Magn. Reson. Imaging* **2009**, *29*, 1008–1012. [CrossRef]

7. Borthakur, A.; Sochor, M.; Davatzikos, C.; Trojanowski, J.Q.; Clark, C.M. T1rho MRI of Alzheimer's disease. *Neuroimage* **2008**, *41*, 1199–1205. [CrossRef]

8. Mangia, S.; Svatkova, A.; Mascali, D.; Nissi, M.J.; Burton, P.C.; Bednarik, P.; Auerbach, E.J.; Giove, F.; Eberly, L.E.; Howell, M.J.; et al. Multi-modal Brain MRI in Subjects with PD and iRBD. *Front. Neurosci.* **2017**, *11*, 709. [CrossRef]

9. Nestrasil, I.; Michaeli, S.; Liimatainen, T.; Rydeen, C.E.; Kotz, C.M.; Nixon, J.P.; Hanson, T.; Tuite, P.J. T1rho and T2rho MRI in the evaluation of Parkinson's disease. *J. Neurol.* **2010**, *257*, 964–968. [CrossRef]

10. Johnson, C.P.; Follmer, R.L.; Oguz, I.; Warren, L.A.; Christensen, G.E.; Fiedorowicz, J.G.; Magnotta, V.A.; Wemmie, J.A. Brain abnormalities in bipolar disorder detected by quantitative T1rho mapping. *Mol. Psychiatry* **2015**, *20*, 201–206. [CrossRef]

11. Mangia, S.; Carpenter, A.F.; Tyan, A.E.; Eberly, L.E.; Garwood, M.; Michaeli, S. Magnetization transfer and adiabatic T1rho MRI reveal abnormalities in normal-appearing white matter of subjects with multiple sclerosis. *Mult. Scler.* **2014**, *20*, 1066–1073. [CrossRef] [PubMed]

12. Wassef, S.N.; Wemmie, J.; Johnson, C.P.; Johnson, H.; Paulsen, J.S.; Long, J.D.; Magnotta, V.A. T1rho imaging in premanifest Huntington disease reveals changes associated with disease progression. *Mov. Disord.* **2015**, *30*, 1107–1114. [CrossRef] [PubMed]

13. Moser, A.D.; Epping, E.; Espe-Pfeifer, P.; Martin, E.; Zhorne, L.; Mathews, K.; Nance, M.; Hudgell, D.; Quarrell, O.; Nopoulos, P. A survey-based study identifies common but unrecognized symptoms in a large series of juvenile Huntington's disease. *Neurodegener. Dis. Manag.* **2017**, *7*, 307–315. [CrossRef] [PubMed]

14. Tereshchenko, A.; McHugh, M.; Lee, J.K.; Gonzalez-Alegre, P.; Crane, K.; Dawson, J.; Nopoulos, P. Abnormal Weight and Body Mass Index in Children with Juvenile Huntington's Disease. *J. Huntingt. Dis.* **2015**, *4*, 231–238. [CrossRef]

15. van der Plas, E.; Langbehn, D.R.; Conrad, A.L.; Koscik, T.R.; Tereshchenko, A.; Epping, E.A.; Magnotta, V.A.; Nopoulos, P.C. Abnormal brain development in child and adolescent carriers of mutant *huntingtin*. *Neurology* **2019**, *93*, e1021–e1030. [CrossRef]

16. Huntington Study Group. Unified Huntington's Disease Rating Scale: Reliability and consistency. *Mov. Disord.* **1996**, *11*, 136–142. [CrossRef]

17. Horton, M.C.; Nopoulos, P.; Nance, M.; Landwehrmyer, G.B.; Barker, R.A.; Squitieri, F.; REGISTRY Investigators of the European Huntington's Disease Network; Burgunder, J.M.; Quarrell, O. Assessment of the Performance of a Modified Motor Scale as Applied to Juvenile Onset Huntington's Disease. *J. Huntingt. Dis.* **2019**, *8*, 181–193. [CrossRef]

18. Tabrizi, S.J.; Langbehn, D.R.; Leavitt, B.R.; Roos, R.A.; Durr, A.; Craufurd, D.; Kennard, C.; Hicks, S.L.; Fox, N.C.; Scahill, R.I.; et al. Biological and clinical manifestations of Huntington's disease in the longitudinal TRACK-HD study: Cross-sectional analysis of baseline data. *Lancet Neurol.* **2009**, *8*, 791–801. [CrossRef]

19. Pierson, R.; Johnson, H.; Harris, G.; Keefe, H.; Paulsen, J.S.; Andreasen, N.C.; Magnotta, V.A. Fully automated analysis using BRAINS: AutoWorkup. *Neuroimage* **2011**, *54*, 328–336. [CrossRef]

20. Powell, S.; Magnotta, V.A.; Johnson, H.; Jammalamadaka, V.K.; Pierson, R.; Andreasen, N.C. Registration and machine learning-based automated segmentation of subcortical and cerebellar brain structures. *Neuroimage* **2008**, *39*, 238–247. [CrossRef]

21. Tustison, N.J.; Cook, P.A.; Klein, A.; Song, G.; Das, S.R.; Duda, J.T.; Kandel, B.M.; van Strien, N.; Stone, J.R.; Gee, J.C.; et al. Large-scale evaluation of ANTs and FreeSurfer cortical thickness measurements. *Neuroimage* **2014**, *99*, 166–179. [CrossRef] [PubMed]

22. Yoo, T.S.; Metaxas, D.N. Open science—combining open data and open source software: Medical image analysis with the Insight Toolkit. *Med. Image Anal.* **2005**, *9*, 503–506. [CrossRef] [PubMed]

23. Bigler, D.; Meadowcraft, M.; Sun, X.; Vesek, J.; Dresner, A.; Smith, M.; Yang, Q. MR Parameter Map Suite: ITK Classes for Calculating Magnetic Resonance T2 and T1 Parameter Maps. *Insight J.* **2008**, *2008*, 237.

24. Tereshchenko, A.V.; Schultz, J.L.; Bruss, J.E.; Magnotta, V.A.; Epping, E.A.; Nopoulos, P.C. Abnormal development of cerebellar-striatal circuitry in Huntington disease. *Neurology* **2020**, *94*, e1908–e1915. [CrossRef]

25. Cronin, T.; Rosser, A.; Massey, T. Clinical Presentation and Features of Juvenile-Onset Huntington's Disease: A Systematic Review. *J. Huntingt. Dis.* **2019**, *8*, 171–179. [CrossRef]

Article

Clinical Manifestation of Juvenile and Pediatric HD Patients: A Retrospective Case Series

Jannis Achenbach [1,*,†], Charlotte Thiels [2,†], Thomas Lücke [2] and Carsten Saft [1]

[1] Department of Neurology, Huntington Centre North Rhine-Westphalia, St. Josef-Hospital Bochum, Ruhr-University Bochum, 44791 Bochum, Germany; carsten.saft@rub.de

[2] Department of Neuropaediatrics and Social Paediatrics, University Children's Hospital, Ruhr-University Bochum, 44791 Bochum, Germany; charlotte.thiels@rub.de (C.T.); luecke.thomas@ruhr-uni-bochum.de (T.L.)

* Correspondence: jannis.achenbach@rub.de

† These two authors contribute to this paper equally.

Received: 30 April 2020; Accepted: 1 June 2020; Published: 3 June 2020

Abstract: Background: Studies on the clinical manifestation and course of disease in children suffering from Huntington's disease (HD) are rare. Case reports of juvenile HD (onset ≤ 20 years) describe heterogeneous motoric and non-motoric symptoms, often accompanied with a delay in diagnosis. We aimed to describe this rare group of patients, especially with regard to socio-medical aspects and individual or common treatment strategies. In addition, we differentiated between juvenile and the recently defined pediatric HD population (onset < 18 years). Methods: Out of 2593 individual HD patients treated within the last 25 years in the Huntington Centre, North Rhine-Westphalia (NRW), 32 subjects were analyzed with an early onset younger than 21 years (1.23%, juvenile) and 18 of them younger than 18 years of age (0.69%, pediatric). Results: Beside a high degree of school problems, irritability or aggressive behavior (62.5% of pediatric and 31.2% of juvenile cases), serious problems concerning the social and family background were reported in 25% of the pediatric cohort. This includes an attempted rape and robbery at the age of 12, as problems caused by the affected children, but also alcohol-dependency in a two-year-old induced by a non-HD affected stepfather. A high degree of suicidal attempts and ideations (31.2% in pediatric and 33.3% in juvenile group) was reported, including drinking of solvents, swallowing razor blades or jumping from the fifth floor with following incomplete paraparesis. Beside dopaminergic drugs for treatment of bradykinesia, benzodiazepines and tetrabenazine for treatment of dystonia, cannabinoids, botulinum toxin injection and deep brain stimulation were used for the improvement of movement disorders, clozapine for the treatment of tremor, and dopa-induced hallucinations and zuclopenthixole for the treatment of severe aggressive behavior. Conclusions: Beside abnormalities in behavior from an early age due to HD pathology, children seem to have higher socio-medical problems related to additional burden caused by early affected parents, instable family backgrounds including drug abuse of a parent or multiple changes of partners. Treatment required individualized strategies in many cases.

Keywords: juvenile Huntington's disease; pediatric Huntington's disease; early-onset Huntington's disease; case series

1. Introduction

Huntington's disease (HD) is an autosomal-dominant hereditary neurodegenerative progressive disorder usually with a most common onset at the age of 30–50 years [1,2]. Nevertheless, motor onset can occur at every age and onsets ≤ 20 years of age are traditionally classified as juvenile HD (JHD) [3]. Recent research suggested the redefining of the term of younger HD patients and to use the term "pediatric" instead of juvenile Huntington's disease for those younger than 18 years, since the definition of juvenile HD (JHD) was found to be blurred and used in different ways [4]. The onset is thereby defined as the presence of unequivocal clinical motor signs (>99% confidence with a diagnostic confidence level (DCL) of four on the "Unified Huntington's Disease Rating Scale" (UHDRS) [5]) caused by HD. Especially in younger patients, motor symptoms may present typically with bradykinesia, dystonia, but also with myoclonus or tremor. The cohort of JHD was described with varieties of motor- and non-motor specific characteristics, thus demands were made for adapting common rating scales, usually to access disease-manifestation in adult HD, for use in children [6].

Case series reports on psychiatric and cognitive nonspecific deficits often accompanied a misdiagnosis or delays in HD diagnosis. JHD patients having an earlier onset also are described to have a longer delay between symptoms and diagnosis of HD [7]. Although the detectable genetic cause with an expansion of cytosine- adenine- guanine (CAG)-trinucleotide repeats in the huntingtin gene (HTT) on chromosome four is obvious to access [1,8], repeat expansions higher than 60–70 CAGs are described as causing a juvenile and more bradykinetic HD phenotype and were frequently detected in juvenile cases described in earlier research [7–10]. Although the genetic cause is unequivocal, characteristics of the clinical disease manifestation, especially in children, are manifold. HD is described as a complex disease with heterogeneous challenges and progressive loss of dependency and increasing disability. For the care of HD patients, several researchers recommend multidisciplinary approaches which are required and include the family, social workers, therapists and physicians to maintain quality of life and to decrease psychosocial problems [11]. For pediatric and juvenile HD, the support not only for children but also for parents or even the whole social system including schools, might be helpful. Psychosocial implications in these settings are described as wide ranged. Former research describes the role of the family and caregivers as burdensome due to disappointing experiences with social and health services, the dissatisfaction of being a caregiver, concern about children, and the loss of needed help from the social or family background due to increasing care for the HD patient and social withdrawal [12]. The unaffected family caregiver is thereby described to need the most support, attention, and help [12]. Many important aspects concerning these socio-medical and psychosocial backgrounds in HD are described [13]. Challenging ethical, social, and legal issues for HD patients, health care professionals, and caregivers are caused by the progressive disease [14]. Most published data consider HD patients and their influence on the social environment. However, further research is necessary regarding social aspects of support and the effects of the caregiver's behavior on the disease and the behavior of the patient and how these interactions affect relations in HD families [15].

The prevalence of HD varies between different geographical regions [16]. Systematic reviews indicated global variations, with an overall prevalence of 5.7 per 10,0000 persons, describing a lower incidence in Asia compared to Europe and North America [3,17]. However, more recently a substantial higher prevalence of HD in the UK (12.3 per 10,0000) and also in Germany was reported, with 9.3 per 10,0000 inhabitants using data of four million insured persons, probably caused by more accurate diagnoses and an improved life expectancy [18,19].The small cohort of juvenile HD patients is described with an equivocal estimated prevalence ranging between 1–9.6% of all HD cases in studies and meta-analyses estimating 4.92% to 6% of all HD patients being juvenile [4,20–23].

Although many important aspects about juvenile HD have been described, no case reports about the recently defined pediatric cohort or comparing research of pediatric and juvenile HD in the boundary of typical characteristics in adult HD were described [4]. Describing the different phenotypes of the clinical manifestation in HD and especially early-onset HD is not only important for the investigation of potential underlying and diverse mechanisms of pathophysiology but primarily for the adapting of different symptomatic therapeutic options. Hereby an exact assessment and rating of predominant symptoms can be crucial because the symptomatic therapy needs to be adapted to the type and extent of individual findings and adjusted frequently during the individual course of the disease [24]. Contrary to treatment of chorea, in a more bradykinetic phenotype, as described for JHD, dopamine agonists may be effective [24]. Case reports and small studies report on an improvement after the dopamine agonist pramipexole or after dosing of medication with L-dopa and amantadine [24,25]. Beside these options for symptomatic therapies, there are no disease-modifying/slowing options or therapeutic options with neuroprotective effects available at the moment [26].

The aim of this case report series is to describe the rare group of early-onset patients, especially in regard to socio-medical aspects and individual or common treatment strategies. In addition, we differentiated between juvenile and the recently defined pediatric HD population (onset < 18 years).

2. Patients and Methods

To classify a cohort of the comparatively very rare pediatric and juvenile HD-patients in more detail, we investigated a retrospective analysis of data from our Huntington Centre North Rhine-Westphalia (NRW). Since its establishment in the year 1995, we have had 25 years of clinical and research experience concerning the treatment and care of adult and early-onset patients suffering from HD. A University Children's Hospital for Neuropaediatrics and Social Paediatrics is affiliated as a part of our institution. The affiliation of the department of Neuropaediatrics and Social Paediatrics to the Huntington Centre NRW for the diagnosis and treatment of HD children is well known among German pediatricians as a result of talks held at pediatric congresses and publications in pediatric journals. Moreover, this cooperation is well known among the German patient support organization, further admissions take place after molecular diagnostic testing in the department of human genetics as part of the Huntington Centre NRW. A rooming in together with a parent was possible when treating children as inpatients.

We analyzed data from our internal digital quality management and hospital information system as well as archived medical letters and examination reports. Apart from this, additional data were collected from archived admission books of the outpatient and inpatient clinic and analyzed, (i) to receive information about a prevalence of juvenile and pediatric HD patients in our clinic, and (ii) to especially evaluate clinical patient-related information for presenting case reports and fundamental correlations of the heterogeneous clinical pictures in early-onset HD.

A special focus was set to describe challenging situations concerning the diagnosis and pharmaceutical treatment of affected patients compared to adult patients as well as challenging situations coming from the socio-medical environment of individual patients. Data concerning socio-medical information were anonymized, analyzed, and based on socio-medical anamnesis in the medical reports. No information was available coming from other sources, such as criminal reports or court proceedings. The investigation was confirmed by our local ethics committee (registration number 20-6892-BR) who agreed on the retrospective anonymized data analysis and publication of information coming from our clinical data management system and medical letters.

3. Results

Considering the last 25 years of the Huntington Centre NRW, we identified 2593 individual patients suffering from HD and presenting in our outpatient and inpatient clinic for seeking medical advice or treatment.

Out of these 2593 patients, 32 individuals (Table 1) in total were identified with an early-onset of a manifest disease when younger than 21, which corresponds to a ratio of 1.23%. Children from all over Germany were admitted to the hospital. Dividing the 32 early-onset patients into pediatric and juvenile HD revealed 18 patients who were classified as pediatric (onset < 18 years of age) and 14 more being juvenile (18 < onset < 21 years of age). Therewith, the proportion of pediatric out of all HD patients presenting in our centre was identified as 0.69%.

Table 1. Clinical characteristics and relevant findings of juvenile and/ or pediatric Huntington's disease (HD) cases from the Huntington Centre, North Rhine-Westphalia (NRW).

First Symptoms	Comorbidities	Predominant Challenging Situation with Special View Regarding Medication
1m→ P→80→1.5→9→P		
Change in muscle tone, frequent falls and increasing gait disorder	Epilepsy	Clinical examination (CE): bradykinetic motor-phenotype with increased muscle-tone, rigidity and spastic component, hypokinesia, cognitive decline and dystrophy. Further diagnostics (FD): comprehensive genetic diagnostics with chromosome analysis, analysis of array-based comparative genomic hybridization, analysis of methyl-CPG-binding-protein 2 gene. MRI: Narrowing of the caudate nucleus on both sides and subsequent gene testing with HD diagnosis. Treatment attempts (TA): initially L-dopa/carbidopa attempt because of a suspected tyrosine hydroxylase deficiency. Low-dosed pramipexole retard (bradykinesia) in combination with quetiapine (due to dopa-induced hallucination) lead to an improvement of movement disorder. Clobazam (anxiety- reducing and sleep- inducing) reduced ongoing sleep-disorder and had a positive effect on dystonia and the increased muscle tonus. Levetiracetam used for the treatment of epilepsy. Course of disease (CD): use of dronabinol for palliative medical care with positive effects on muscle tonus, dystonia and body mass index (BMI).
2f→P→73→9→13→P		
Problems in school, pronounced tremor with medium amplitude, gross and fine motoric function		Anamnesis (A): early genetic testing because of positive family history, but retrospective symptoms already exist for the past 4 years. CE: initially hypomimia, lack of arm swinging, cognitive deficits. Socio-medical aspects (SM): re-educated after diagnosis in a special school, which helped enormously. After depression with several suicidal attempts, irritable behavior, lack of impulsion, apathy, schooling no longer possible. Long-term psychiatric ward. FD: electroencephalography (EEG): isolated bi-frontal sharp-wave-like activity and typical epilepsy potentials but no seizures observed while treatment with valproate (for stabilizing irritability). TA: pramipexole (treatment of bradykinesia), venlafaxine (antidepressant), valproate retard, lithium (antidepressant and because of suicidal intent), quetiapine (antipsychotic). In adulthood, worsening of dystonia and spasticity (tiazidine initiated), swallowing and because of improved mood and absence of suicidal tendencies lithium stopped.
3m→P→48→13→14→P		
Problems in school, depression, tremor		A/CE: stressful family situation, father died due to suicide, sister died due to HD. Unspecific symptoms reported (depression, due to HD or family situation?). No genetic testing during the first visit. An additional tremor three months later indicated the genetic testing. FD: diagnostic of cranial MRI, awake, sleep EEG, electrocardiography (ECG) inconspicuous. SM: arrangement with school, compensations for disadvantages helped remaining in class. TA: psychotherapy, occupational therapy but no additional pharmacological therapy necessary at the present. CD: mild bradykinesia in finger tapping without negatively influencing everyday life.

Table 1. *Cont.*

First Symptoms	Comorbidities	Predominant Challenging Situation with Special View Regarding Medication
4m→P→81→Infant age→9→P		
Progressive motoric dysfunction, speech problems and problems in school	Epilepsy Severe expressive speech disorder	A: postnatal difficulties in thriving and swallowing. CE: bradykinesia in addition to hyperkinetic restlessness at night. FD: cranial MRI (second year of life) revealed a reduction of brain volume (retrospective evaluated as physiological expansion of the external cerebrospinal fluid spaces in infancy). TA: pramipexole retard (improved bradykinesia), valproate retard (positive influence on irritability), quetiapine (reduction of irritability and hallucinations), L-dopa/carbidopa (positive influence on rigidity) CD: severe swallowing disorder, but percutaneous endoscopic gastrostomy (PEG) rejected.
5f→P→47→10→16→M		
Attention deficit hyperactivity disorder (ADHD), problems in school, depression	Borderline personality disorder	A: initially presented to human genetic specialists by foster parents because of positive HD family history. CE: fine motor skills problems in initial testing. FD: EEG with abnormal findings, without therapeutic consequences. TA: refrain of therapy with dopamine agonists (e.g., pramipexole) due to optical hallucinations. Psychiatric symptoms more severe and main symptoms, motoric symptoms with slight tremor and rigidity. Low-dose risperidone (reduction of irritability and aggressive behavior), quetiapine, venlafaxine (as an antidepressant with only moderate success) and clozapine (because risperidone was not tolerated in higher doses). Clozapine improved emotional instability and psychosis but also tremor. CD: earlier multiple suicide attempts (taking tablets, drinking solvents and swallowing razor blades). Patient is now an adult with a stable course and stabilized psychiatric symptoms in the course of the disease. Works in a workshop facility, accommodated in residential facility.
6m→P→63→7→13→M		
ADHD, problems in school, behavioral problems		A: initially diagnosed and treated as ADHD by unconnected children psychiatry, massive behavior problems. Before inconspicuous infancy (except for a slight delay in speech development). Methylphenidate (to improve attention problems) without lasting success. Patient then reconsidered as "not trainable". CE: discreet bradykinesia, discreet intention tremor. Fine motoric skill disturbance and increased muscle tone of the lower extremities. Afterward molecular genetic HD diagnosis. FD: pathological EEG without seizures. SM: Aggressive behavior towards other children and adults. Massive crimes documented by police with aggression at the age of six years (documented rape attempt on a girl, robbery on an elderly woman and theft of a handbag in the age of 12) led to in- patient diagnostics in a child and youth psychiatry. Persistent behavioral problems with short attention span, lack of controllability for own behavior and verbal aggression persisted. Diagnosis however led to correction of expectations and consequently to relaxing of the loaded situation. TA: risperidone (improved situation), zuclopenthixole (positive influence on irritability) stabilized conditions.

Table 1. *Cont.*

First Symptoms	Comorbidities	Predominant Challenging Situation with Special View Regarding Medication
7f→P→85→6→9→P		
Motor abnormalities increasingly occurred on the left side	Epilepsy	A/CE: Expressive speech development delay (received speech therapy). Increasing dystonia of the left foot, three years after the first motoric symptoms and following genetic HD testing. Increasing lethargy in the course of disease, dystrophic (weight < 3 percentile, length < 3 percentile). Saccadic eye movement, chameleon tongue, dystonia of left foot, tremor, bradykinesia, increased muscle tone and reflexes. FD: pathological EEG. TA: venlafaxine or tetrabenazine suggested (lethargy). Various pharmaceutical trials (including L-dopa, amantadine, trihexyphenidyl) without success. CD: Tonic- clonic movement considered as a seizure and valproate used as an anticonvulsive. Dronabinol at request of mother improved dystonia.
8m→P→68→15→20→P		
Bradykinesia, progressive loss of concentration		A: positive family anamnesis, but initially no genetic testing wanted by patient and mother. TA: After several frustrating trials with pharmaceutical treatments, Deep brain stimulation (DBS): stereotactic implant of electrons in globus pallidus both sides. An initially setting effect with reduction of dystonia and rigor leading to a better body posture, better fine motor activity and less blepharospasm but worsening of gait and postural instability reported, in summary no significant objective long-term effect. Additional quetiapine, L-dopa/benserazide, pramipexole, seroquel. Fresurbin through PEG. Dopa- induced psychosis (optic and acoustic hallucination) lead to reduction of madopar. CD/SM: Hypersalivations and frequently bronchopneumonia. Organization of a school accompaniment.
9m→P→70→13→17→P		
Progressive movement disorders with hypokinetic-rigid aspects	Motoric axonal polyneuropathy DD: critical illness polyneuropathy	A: initially no genetic testing wanted by patient and mother (brother of case 8). TA/CD: Initially Madopar and memantine for cognitive symptoms afterwards DBS without beneficial effect, PEG with dislocation/ hematin vomiting, hypernatremia/ thrombopenia, respiratory decompensation. Urine, defecation-incontinence. Intermittent beneficial botulinum-toxin injection for treatment of dystonia in upper-extremity. Progressing exanthema (better after stopping of sifrol, Nexium, sirdalud). Amantadine, rotigotine lead to an improvement and better position of contract arms/ head. Due to ill skin loss of water with hypertonic dehydration. Topramax due to dystonia lead to an improvement of tremor. Additionally cannabinoids (Sativex-Spray©) leads to marked improvement of dystonia. SM: organization of a school accompaniment.
10f→J→61→20→23→P		
Psychiatric symptoms with attacks of suffocation and heart-pain without clinical correlate, depression	Sudden infantile death syndrome with reanimation	A: At the age of 23 already lived in a care home without HD diagnosis. Before multiple times inpatient in psychiatry because of panic attacks. Post-traumatic stress disorder (PTSD, no further circumstances described). Reported on unspecific described general pain of the whole body during clinical assessments. CE: stiff movement disorder, dysarthria. TA: Trials with baclofen, lorazepam, madopar, mirtazapine, seroquel, tolperisone, amantadine. L-dopa improved movement disorder with "more fluid". mydocalm (pain) improved sleep, before suffering from nightmares. Stop of amantadine, instead memantine and artane (dystonic posture of the head) lead to a better movement of motoric and psychiatric situation.

Table 1. *Cont.*

First Symptoms	Comorbidities	Predominant Challenging Situation with Special View Regarding Medication
11m→P→>60→8→11→M		
Motoric restlessness and multiple psychiatric problems		A/SM: Since early childhood psycho- social conflict situations and multiple continuous mistreatments due to father who made the patient alcohol- dependent at the age of two. Mistreated with cigarette pushing against the arm and aggressive behavior. Afterwards lived with grandfather (by care). TA: ritalin initially, later risperidone, valproate, tiapride CD: challenging loss of weight, swallowing and chewing difficulties.
12m→P→64→9→11→P		
Aggressive behavior and hyper movement of eyes	Epilepsy Amblyopia	A: Loss of concentration, difficulties finding words and loss of memory. A described reduction of intelligence. FD: MRI with bilateral signal increase Globus pallidus calcification and symmetric expansion of the lateral ventricle. SM: psycho-social conflict situation in family with massive problems: traumatic situations (saw brother drowning, diverse partners of mother mistreated the patient and mother). TA: valproate (Convulex) for epilepsy resulted also in less agitation and aggressive behavior after risperidone. CD: swallowing and chewing difficulties.
13m→P→Extended CAG, without reporting of exact repeat length→14→17→P		
Changing of speech and bradykinesia		A/CE: aggressive behavior and bradykinetic rigid movement disorder. FD: computed tomography (CT) with atrophy of Caput nuclei caudate and expansion of lateral ventricle front horns. SM: increased alcohol abuse since the age of 18, marihuana abuse (treated around the age of 15–17), worked in supervised workshop. TA: Rotigotine- pavement lead to improvement of movement. Bupropion (elontril) lead to an improvement of impulsive behavior, additionally Seroquel.
14m→P→Extended CAG, without reporting of exact repeat length→6→8→M		
Conflicts and aggressive behavior		CE: bradykinesia with worsening of movements. SM: Lived in a boarding school, before in a care family, aggressive behavior with impulsive outbursts. Nicotine abuse. TA: ebixia, lamictal, cipramil and requip (as dopaminergic stimulation) lead to an improvement of symptoms.
15f→J→52→19→23→M		
Dystonic-hypokinetic symptoms, loss of concentration	Bronchial asthma, atopic dermatitis	A/CE: Worsening of fine motor skills (writing) and loss of concentration. Drug abuse until the age of 23 (marihuana, PEP, cocaine), intermittent alcohol abuse and nicotine abuse. Stopped clinical evaluation and inpatient therapy herself because she already felt an improvement of symptoms. TA: Requip modutap lead to a better gang picture and better fine motoric skills. Did not take medication further. Second inpatient evaluation, again discharge against medical advice (she did not see an improvement). Berotec inhaler (if needed for distress), novaminsulfon (due to chronic pain), trial with tetrabenazine stopped due to a described "Heart pain".

Table 1. *Cont.*

First Symptoms	Comorbidities	Predominant Challenging Situation with Special View Regarding Medication
16m→J→Extended CAG, without reporting of exact repeat length→20→25→P		
Hyperkinesia motoric clumsiness	Drug-induced parkinsonism	A/CE: Molecular genetic testing due to hyperkinesia. Description of a rapidly massive worsening of chorea during 3-4 weeks. TA: Induced increasing of tiapride, swallowing problems, fatigue and no improvement of chorea, more a subjective worsening and parkinsonism. Because of the side effect and expected same side effects of other typical postsynaptic neuroleptics: Therapeutic target with tetrabenazine which improved symptoms very well (tiapride, tetrabenazine and trihexyphenidyl in combination) with three further stable years. CD/TA: at the age of 28 more dystonia and myocloniform hyperkinesia lead to a trial with valproate and reduced tiapride, tetrabenazine with satisfying improving of symptoms.
17f→J→Extended CAG, without reporting of exact repeat length→20→23→unknown, no HD family history		
Agitation, cognitive deficits	Emotionally unstable personality disorder	A/CE: Speech problems, swallowing problems, movement disorder in terms of gear insecurity, problems with coordinative movements. SM: Reporting on conflicts concerning the social background, lack of care. Aggressive and oppositional behavior, depression with suicidal ideation. Went to a special school, but without school-leaving certificate and lived in a residential facility because of integration problems. TA: Initial treatment with sulpiride, discontinuing of medication lead to an improvement. Lamictal induced because of fluctuation of affliction, amantadine lead to visual hallucination, theophylline, Broncho spray, Belladonysat additional.
18m→J→55→20→23→M		
Tremor		A/CE: Initially tremor, described by parents and patient. Intermittent suicidal thoughts, aggressive and irritable behavior. Posture and action tremor in clinical assessment, also tremor of the tongue. Only very slight intermittent hyperkinesia, more bradykinesia and rigid tone. TA: citalopram in combination with quetiapine and dopamine-agonist pramipexole initially in small doses due to danger of psychotic disorder lead to an improvement.
19f→J→53→20→22→P		
Motoric clumsiness	Epilepsy Hypothyroidism	A: massive psychiatric symptoms with depression, irritability, aggressive behavior, perseverative/obsessive behavior. FD: EEG with evidence of spike-wave-complexes left parietal. TA: Difficult pharmacological situation due to recurring unrest and fatigue: Trials with L-thyrox, levetiracetam, mirtazapine, nitoman, lamotrigine, zuclopenthixole (because of affective disorder and irritability). Challenging BOL hyperkinesia (improved after tetrabenazine), reduction of levetiracetam improved fatigue. Pharmaceutical trials with tiapride, sulpiride, gabapentin, pipamperone. CD/TA: Dysarthria, suicidal attempt, jumped from fifth floor with incomplete paraparesis. More myoclonic movement disorder in upper extremities lead to a treatment with valproate.

Table 1. *Cont.*

First Symptoms	Comorbidities	Predominant Challenging Situation with Special View Regarding Medication
20f→P→n.d.→n. d.→16→P		
n.d.	Epilepsy Dystonic tetra paresis Anemia	CE: dystonia, hypokinetic movement disorder with spasticity. TA: Madopar lead initially to worsening, medication was stopped. Lioseral against spasticity lead to an improvement of symptoms. CD/TA: Tremor, increasing of creatine kinase and fever lead to the diagnosing of a malignant neuroleptic syndrome (MNS) which was stable after discontinuing the medication. Increase of spasticity and psychotic behavior lead to trials with tavor, cipramil, lioresal, seroquel, zuclopenthixole. Before-massive psycho-motoric unrest with attacks of screaming. Intermittent infections (bronchopulmonary due to aspiration?). Trials with levodopa, oxazepam, diazepam, keppra, nexium, baclofen, seroquel, tolperisone, durogenesic patch, tizanidine. As needed: tavor (fear and restlessness), melperon (restlessness), diazepam (spasticity or seizure).
21f→J→60→20→22→P		
Psychiatric symptoms, depression. Hyperkinesia	Sterilization	CE: only slight hyperkinesia initially (trunk, extremities). FD: MRI with extension of the lateral ventricles and lateral ventricle anterior horns. CD: Massive loss of weight (20 kg in two years) and recurring infections. Changing of sleep- awake rhythm (trial with mirtazapine improved situation). Tiapride and tetrabenazine stabilized motoric and psychiatric situation.
22m→P→56→17→20→M		
Hyperkinesia and motoric clumsiness, depression	Thoracicos-teochondrosis	CE: Hyperkinesia and a combination of dystonic components as part of the movement disorder. SM: finished school and professional training. FD: MRI with subcortical atrophy. CD/TA: Suicidal ideation (driving against a tree), self-harming behavior and sleep disorder. Mirtazapine, tiapride improved situation greatly. Increasing aggressive behavior lead to trials with citalopram, valproate, risperidone, arcoxia, memantine, zopiclone, pipamperone, elontril, tavor. Because of BOL- hyperkinesia botox was induced into the masseter. Disclaiming of food intake, massive perseverative behavior and decubitus during hospitalization. Changing of day/night rhythm = better after reduction of clozapine and increase of Haloperidol. Urine, defecation- incontinence.
23m→J→57→20→23→P		
Restlessness, worsening of fine motoric skills		CE/TA: Loss of concentration plus senso-motoric dysarthria. Hypokinetic- rigid symptoms with good response of levodopa, long lasting dopamine-agonist starting with pramipexole ret. CD/TA: Worsening of bradykinesia, tendency to fall. Trial with amantadine, quetiapine because of a sleep disorder, zopiclone as needed.
24f→J→66→19→21→unknown, no HD family history		
Hyperkinesia and restlessness.	Epilepsy with grand mal seizure	CE: Initially ocular interrupted pursuits, increased saccade latency and slow velocity. Fine motoric movement disorders with finger taps/ pronate supinate hands severe slowing and irregular. TA: Tetrabenazine even before diagnosis of HD, baclofen for dystonia. Diazepam for treatment of epilepsy. CD: suffered from slight intermittent chorea in combination with marked/ prolonged dystonia especially in upper extremities and markedly slow bradykinesia

Table 1. *Cont.*

First Symptoms	Comorbidities	Predominant Challenging Situation with Special View Regarding Medication
25m→P→64→15→16→P		
Psychiatric symptoms	Allergic asthma	A/CE: Violent/aggressive behavior, apathy, perseverative, obsessive behavior. Chorea only slight/intermittent BOL/Trunk. Fine motoric skills difficulties, saccades and ocular pursuit. Previous suicidal attempts. TA: escitalopram (obsessive-compulsion) and quetiapine (mood stabilization) improved situation.
26m→P→67→9→11→n.d.		
n.d.	n.d.	n.d.
27f→J→58→20→n.d.→n.d.		
n.d.	n.d.	n.d.
28m→J→67→18→n.d.→n.d.		
n.d.	n.d.	n.d.
29m→J→53→18→n.d.→n.d.		
n.d.	n.d.	n.d.
30f→J→50→18→n.d.→n.d.		
n.d.	n.d.	n.d.
31f→P→77→12→n.d.→n.d.		
n.d.	n.d.	n.d.
32m→J→n.d.→18→n.d.→n.d.		
n.d.	n.d.	n.d.

Patients are itemised to: individual patient number (numeric) and female (f) or male (m) sex, pediatric (P) or juvenile (J) onset, CAG repeat- expansion length, age of onset (years), age of diagnosis (years), paternal (P) or maternal (M) inheritance. Description of predominant challenging situation classified according to anamnesis (A), clinical examination (CE), further diagnostics (FD), socio-medical aspects (SM), treatment attempt (TA) and course of disease (CD). Abbreviation n.d. (no given data).

Considering the described cohort of early-onset patients, 59.4% appeared to be male patients. In the pediatric cohort, the rate of male patients (72.2%), corresponding to 13 out of 18 patients, was higher than in the juvenile cohort, where a lower rate of 42.9% (six out of 14 patients) appeared to be male.

Analyzing the early-onset cohort more closely revealed, 16 cases with a paternal inheritance model, seven cases with a maternal inheritance, two cases without any HD in the anamnesis of the accompanied parents or rest of the family anamnesis, and seven cases without given data from family anamnesis. As a reason for missing family history, detailed information from medical reports were not available any more in seven cases because reports were destroyed after the minimum of 10 years duty of archiving. The 23 cases with reported data from the family anamnesis therewith revealed a paternal inheritance of 69.6% in the monitored cohort.

The reported CAG repeat length of all early-onset patients ($n = 25$) revealed a median expanded allele length of 63. After diving into the different groups of pediatric ($n = 14$; M = 67.3; SD = 11.6) compared to juvenile HD patients ($n = 11$; M = 57.4; SD = 5.6), independent *t*-test (IBM® SPSS® Statistic V25, IBM Corp., Armonk, NY, USA) demonstrated significantly higher CAG repeats in the pediatric group ($t(19.6) = -2,8$, $p = 0.011$).

In the juvenile cohort, a mean age of onset ($n = 14$) was identified as 19.3 years (SD 0.9) and the age of diagnosis ($n = 9$) on average identified as 23.2 years (SD 1.2) (Table 2). The pediatric cohort revealed a mean age of onset ($n = 16$) of 10.3 years (SD 4.1) and an average age of diagnosis ($n = 17$) of 13.5 years (SD 3.8). Describing the cumulated percentages in the pediatric group revealed that 50% of the cohort had an onset ≤ 9 years of age and 75% of the cohort ≤ 13 years of age. Of the pediatric group, 52.9% was diagnosed with an age ≤ 13 years and 76.5% with an age ≤ 16 years.

Table 2. Demographics of the pediatric and juvenile HD cohort.

Demographics	Pediatric (*N* = 18)	Juvenile (*N* = 14)
Male/female sex (%)	72.2/27.8	42.9/57.1
CAG- repeat-expansion length	67.3 (SD 11.6, $n = 14$)	57.4, SD 5.6, $n = 11$)
Age of onset (AO) (years)	10.3 (SD 4.1, $n = 16$)	19.3 (SD 0.9, $n = 14$)
Age of diagnosis (AD) (years)	13.5 (SD 3.8, $n = 17$)	23.2 (SD 1.2, $n = 9$)
Mean time between AO and AD (years)	3.5 (SD 1.9, $n = 15$)	3.0 (SD 1.0, $n = 9$)
Paternal/maternal inheritance	68.8/ 31.2 ($n = 16$)	71.4/28.6 ($n = 7$)

Comparing these median reported ages in juvenile and pediatric groups using independent t-tests demonstrated a significantly earlier onset ($t(16.6) = 8.4$, $p \leq 0.001$) and diagnosis ($t(21.4) = 9.1$, $p \leq 0.001$) in the pediatric group.

The mean times measured in years between age of onset and diagnosis revealed no significant difference between the pediatric ($n = 15$; M = 3.5; SD 1.9) and the juvenile group ($n = 9$; M = 3.0; SD 1.0) in the independent *t*-test ($t(22) = -0.8$, $p = 0.411$).

The motoric components of the disease manifestation revealed especially hypokinetic-bradykinetic compounds and rigidity being present in the majority of the pediatric cohort (reported in 11 out of 16 cases: 1, 2, 3, 5, 6, 7, 8, 9, 13, 14, 20) and also present in five out of nine cases with juvenile HD (cases 15, 18, 20, 23,24).

Six out of 16 pediatric cases (case 1, 2, 7, 8, 9, 20) revealed the presence of dystonia and therewith in 37.5% of the pediatric collective compared to 22.2% of juvenile cases (two of nine cases; cases 16, 24).

Hyperkinetic movement disorders in terms of chorea were present in cases 16, 18, 19, 21, 24 (all juvenile) and slightly in cases 22, 24 (both pediatric), which represents 55.6% of the juvenile and 12.5% of the described pediatric cohort. Another hyperkinetic movement disorder, myoclonus, was identified in two out of nine juvenile phenotypes (22.2% of the juvenile cohort).

Out of 16 pediatric patients, we identified six patients who were diagnosed with epilepsy, which corresponds to 37.5% of that pediatric cohort. In comparison, we identified two out of nine patients with epilepsy in the juvenile cohort (22.2% of the cohort).

Six out of 16 pediatric patients presented with tremor in their clinical assessments, which corresponds to 37.5% of the cohort. In the juvenile cohort one patient suffered from tremor (11.1% of the cohort).

Changing and difficulties of speech were present in cases 4, 6, 7, 13 (all pediatric) and case 17 (juvenile), and therewith relevant in 25% of the pediatric and 11.1% of the juvenile cohort.

Concerning the diversity of psychiatric problems, aggression and irritability was reported in 10 out of 16 pediatric cases (cases 2, 4, 5, 6, 11, 12, 13, 14, 22, 25), which corresponds to a ratio of 62.5%, and in three juvenile cases (cases 17, 18, 19), which is a ratio of 33.3% in the juvenile cohort. Suicidal ideations or attempts were described in eight cases in total, whereby five cases were pediatric (cases 2, 3, 5, 18, 22) and three cases were juvenile (cases 15, 17, 19), which reports a ratio of 31.2% in the pediatric and 33.3% in the juvenile group.

The presence of apathy in cases 2, 25 (both pediatric) as well as obsessive behavior in case 19 (juvenile) and 25 (pediatric) were described as other psychiatric symptoms.

The main aspects of social pediatric characteristics especially included problems in school, described in five pediatric cases (cases 2, 3, 4, 5, 6) and one juvenile case (case 7) as frequent initial symptoms, especially in the pediatric group and also accompanied with a cognitive decline (cases 1, 2, 9, 17).

Furthermore, other patients showed criminal behavior characteristics (case 6) as well as substantial problems with alcohol and drug abuse (cases 11, 13, 15) and serious problems concerning the social background and family, with violation and conflicts in the family (all pediatric cases 3, 11, 12, 14), which equates a ratio of 25% in the pediatric cohort.

Considering diagnostic interventions, five out of 16 pediatric patients (31.2%) and one juvenile case (21) had an MRI, whereby only in one case was an inconspicuous result reported.

4. Discussion

Juvenile HD is described as a very rare manifestation of HD with an estimated prevalence of 1–9.6% of all HD cases in studies and meta-analyses estimate 4.92% to 6% of all HD patients being juvenile [4,12–15]. Out of 2593 individual HD patients treated within the last 25 years in the Huntington Centre NRW, 32 subjects were analyzed with an early onset younger than 21 years. This relates to a proportion of 1.23%. This is within the range of published data. Several problems arising from the terminology JHD are currently being discussed. Many patients seem to be called juvenile because of their more bradykinetic phenotype even if they are older than 21 [21,27,28]. Thus, in JHD the bradykinetic phenotype can be seen as being most characteristic, which is usually associated with a more dopaminergic therapy. However, there seems to be a large cohort of HD patients who are already adults suffering from a more predominant bradykinetic-rigid clinical phenotype, called JHD. Some of them perhaps even had a motor onset beyond 20 years of age and are called JHD only because of their more bradykinetic symptoms [29]. This might explain the high range for estimation of prevalence of 1–9.6% and also explain why the prevalence in our cohort does not correspond with the higher mean described above and classified in the lower range since we used the strict definition of an age below 21 year for our collective. This difference is even more obvious if the more recently definition of pediatric HD is used, with only 0.69% of all cases. This remarkably low rate might be caused due to fewer admissions of early-onset HD cases to our centre. However, as part of our institution, a university children's hospital for Neuropaediatrics and Social Paediatrics is affiliated and well known by the German pediatricians society and patient support organisation. Since to the best of our knowledge, this is the only known cooperation like this in Germany and the identified cases were submitted nationwide, we would expect a bias to even more JHD cases and not less in our cohort.

As a first aspect to discuss in our case report series we detected several findings concerning socio-medical aspects not only caused by HD pathology but also caused by early affected parents, instable family backgrounds, including drug abuse by a parent or multiple changes of partners. The results of our research are in line with earlier findings. Massive burdens for caregivers, HD families and especially for the patient arise from socio-medical problems in an observed mutual interplay of the social surrounding and the affected person [12].

That implements not only the pharmacological treatment of patients but also more socio-medical supports like the organization of a school accompaniment, support by a social worker and support of the non- affected parent including psychotherapy in order to reach and maintain a certain degree of education. Especially, problems in school were often described as initial symptoms in many cases, they were also accompanied by cognitive decline. In this context, the establishing of the diagnosis HD was extremely helpful for most cases in our experience since it was possible to reduce learning pressure in school and define new targets for education in discussion with the school. Children in most cases were not depressed or burdened by the establishing of the HD diagnosis, but relieved by finding an explanation for their difficulties at school and by defining new, more social aspects of participating in school visits. This is even more important since a delay in diagnosis is described in our collective and also published data in many cases [30] as one reason for late diagnosis the resistance of parents was reported. Following our experience, we would rather recommend an earlier diagnosis than a delay in diagnosis to parents in order to reduce the burdens of their children in social-medical aspects. As another reason for late diagnosis, a lack of knowledge among health care professionals is discussed, possibly also caused by unsuspectingness about the heterogeneous motoric symptomatic with a more bradykinetic motor phenotype in early- onset HD. In many of our cases, expensive and stressful or even painful diagnostic interventions like lumbar punctures were performed even if family history of HD was positive. In addition, some cases were initially misdiagnosed and mistreated as attention deficit hyperactivity disorder or borderline disease. Thus, we can support former described findings of delayed diagnosis in this subgroup [30]. From our perspective it is decisive to enable HD patients and families a continuous care through a specialized centre. For this purpose, participating in a prospective registry study (e.g., ENROLL-HD) might enable annual visits, review of socio-medical aspects, and the family anamnesis. If an affected HD family member or caregiver is in contact with a specialized centre and reports difficulties with a child, an investigation of this child should be offered. Additionally, index patients should be asked specifically about their own children and whether there are any abnormalities observed. A multidisciplinary setting including an early involvement of neurologists and neuropaediatricians is crucial and extremely helpful in our view. If there are typical HD symptoms, in most cases being investigated in a centre, diagnosis is not substantially delayed in our experience. However, if children are not initially investigated in a centre, we observed long delays and multiple unnecessary and stressful examinations including lumbar puncture and others in many cases. This might be due to several reasons: the lack of knowledge about more bradykinetic symptoms, concerns regarding performing genetic testing in a child with a lack of knowledge about the differential diagnostic-testing procedure, or concerns to make such a severe diagnosis as HD. Thus, the only way to avoid delays in diagnosis in these cases might be through the participation of professionals from a centre in meetings of patient-organisations, participation in congresses for professionals (e.g., neurologists or neuropaediatricians), public relations work about HD, and the offer of training courses for the specialized public. More important, from our experience, when making the diagnosis of HD in children is that almost all children were more relieved by the diagnosis than burdened. Regardless of the therapy with drugs, most children had a relief if their existing problems in school, in sports or with friends could be explained due to the HD diagnose. For example, subsequent changes or help in school setting reduced stress and declined fear of failure. Our impression is, that children, as in other severe diagnoses like cancer, seem to cope with the diagnosis well in most cases whereas diagnosis is often more difficult for the parents.

As a second aspect, it is important to distinguish between the heterogeneous movement disorders in early- onset HD, with bradykinesia, dystonia, tremor or myoclonic hyperkinesia in order to achieve a beneficial pharmacological treatment [31–33]. Many of our cases were treated with dopaminergic drugs or amantadine for an improvement of bradykinesia, benzodiazepines or tetrabenazine was used for treatment of dystonia with beneficial effects. In single severe cases, cannabinoids and deep brain stimulation (DBS) was used for more generalized dystonia or botulinum toxin- injection for focal dystonia. For DBS only, intermittent positive effects and only mild beneficial effects after treatment with trihexyphenidyl were observed [33]. A relatively high amount of 37.5% of the pediatric cohort and 11.1% of the juvenile cohort, respectively, suffered from tremor. In most cases tremor was treated with dopa-agonists like pramipexole, in one case an excellent benefit was observed after treatment with clozapine with an initial indication of dopa-induced hallucination [34]. Valproic acid was used for treatment of epilepsy, as a mood stabilizer but also for treatment of myoclonic hyperkinesia in two cases [31]. As listed in our reports, in many cases a combination of medication or changes in medication in the course of the disease was necessary.

The occurrence of epilepsy with seizures in JHD is described as an additional important clinical feature [7,10,35]. Although not much is known about why epilepsy occurs more in JHD than in adult-onset HD, retrospective studies report on a very frequent occurrence of 38% in a collective of juvenile HD patients, which corresponds to a rate of 37.5% in our collective [36]. Valproic acid was used for treatment in many cases but also levetiracetam, lamotrigine and benzodiazepines. No case of increased irritability as a known side-effect was observed regarding treatment with levetiracetam.

As another clinical feature, a delay of speech development was observed in five cases and therewith present in 25% of the pediatric cohort. Remarkably, the speech development delay and a consequent logopaedic therapeutic trial was even described in advance of other clinical symptoms or diagnosis of HD. As family anamnesis revealed a younger and older brother without any suspected speech development delay, we assumed a potential organic cause due to HD.

Finally, we observed a very high amount of different psychiatric disturbances like aggression and irritability in 62.5% of the pediatric and 33.3% of the juvenile cohort, respectively. Substantial aggressive and criminal behavior causing serious problems for the family and social network as well as for the patient was observed in case 6 in particular. His mother was suffering from HD and already severely affected at the time of first admission. Besides the potential organic burdens due to the disease, we observed challenging situations for the patient constituted by a socially disadvantaged family situation possibly leading to delinquent behavior due to missing corrective behavior of the father. He was burdened also as the caregiver for the affected mother during that time. Disease-related missing impulse control, an additional impact of the social surrounding, or even further developments of the common puberty-age with a personal-development process can be discussed as being decisive for his behavior. Remarkably, older and younger siblings of this case did not show any delinquent behavior, that is why we assume the described multifactorial reasons including HD related brain changes and not solely the influence of his social background on the described behavior. Suicidal ideations or attempts were described in 31.2% of the pediatric and 33.3% of the juvenile cohort, as well as other psychiatric symptoms, with apathy in two pediatric cases and obsessive behavior in one juvenile and pediatric case [37]. Treatment was effective for most of these cases following guidelines and common psychiatric treatment strategies [27]. Serotonin-selective-reuptake inhibitors (SSRI) and mirtazapine, especially if sleep problems occurred, were the most common medication for treatment of depression. Quetiapine and risperidone were used for the treatment of irritability and aggressive behavior, whereas in single cases zuclopenthixole for the treatment of severe aggressive behavior was also effective [38]. As potential side- effects, bradykinesia and rigidity worsened especially after treatment with risperidone. As a relevant side-effect of therapies with L-dopa or dopa- agonists, hallucinations, especially optical hallucinations, might occur. In five of our 25 cases with complete records, optical, and in one case acoustic, hallucinations were caused due to the dopaminergic treatment and improved after reduction of dosage or with additional neuroleptic treatment [34].

As a limitation, our case reports are based on a retrospective analysis of medical reports and not on standardized implemented scales for accessing of symptoms. Moreover, in seven cases, no additional detailed medical reports were available anymore. However, this more descriptive research approach enabled the depiction of heterogeneous and manifold aspects in early-onset HD which might not be captured in standardized scales, such as the described socio-medical aspects. More in general this is a very rare subgroup of HD, which might limit power for further research in many aspects.

5. Conclusions

In summary, beside from early abnormalities in behavior due to HD pathology, children seem to have higher socio-medical problems related to additional burden caused by early affected parents and instable family backgrounds.

Author Contributions: J.A.: Conception, Organization, Execution, Design, Writing of the first draft; C.T.: Conception, Organization, Execution, Design; T.L.: Conception, Review and Critique; C.S.: Conception, Organization, Design, Review and Critique. All authors have read and agreed to the published version of the manuscript.

Acknowledgments: We acknowledge support by the DFG Open Access Publication Funds of the Ruhr-Universität Bochum.

Conflicts of Interest: Saft reports personal fees/honoraria from Teva Pharma GmbH, as well as non-financial support and other support from ENROLL-HD study (CHDI), PRIDE-HD (TEVA), LEGATO (TEVA), and Amaryllis (Pfizer), ASO (IONIS Pharmaceuticals and Roche AG) for the conducting of studies and grants from Biogen all outside the submitted work and without relevance to the manuscript. The other authors declare they have no conflict of interest concerning the research related to this manuscript.

References

1. Walker, F. Huntington's disease. *Lancet* **2007**, *369*, 218–228. [CrossRef]
2. Roos, R.A. Huntington's disease: A clinical review. *Orphanet J. Rare Dis.* **2010**, *5*, 40. [CrossRef]
3. Quarrell, O.; Nance, M.A.; Nopoulos, P.; Paulsen, J.S.; Smith, J.A.; Squitieri, F. Managing juvenile Huntington's disease. *Neurodegener. Dis. Manag.* **2013**, *3*, 267–276. [CrossRef]
4. Quarrell, O.; Nance, M.A.; Nopoulos, P.; Reilmann, R.; Oosterloo, M.; Tabrizi, S.J.; Furby, H.; Saft, C.; Roos, R.A.C.; Squitieri, F.; et al. Defining pediatric huntington disease: Time to abandon the term Juvenile Huntington Disease? *Mov. Disord.* **2019**, *34*, 584–585. [CrossRef]
5. Kieburtz, K.; Penney, J.B.; Corno, P.; Ranen, N.; Shoulson, I.; Feigin, A.; Abwender, D.; Timothy Greenarnyre, J.; Higgins, D.; Marshall, F.J.; et al. Unified Huntington's disease rating scale: Reliability and consistency. *Mov. Disord.* **1996**, *11*, 136–142. [CrossRef]
6. Cronin, T.; Rosser, A.; Massey, T. Clinical Presentation and Features of Juvenile-Onset Huntington's Disease: A Systematic Review. *J. Huntingt. Dis.* **2019**, *8*, 171–179. [CrossRef]
7. Gonzalez-Alegre, P.; Afifi, A.K. Clinical characteristics of childhood-onset (juvenile) Huntington disease: Report of 12 patients and review of the literature. *J. Child Neurol.* **2006**, *21*, 223–229. [CrossRef]
8. Ribai, P.; Nguyen, K.; Hahn-Barma, V.; Gourfinkel-An, I.; Vidailhet, M.; Legout, A.; Dodé, C.; Brice, A.; Dürr, A. Psychiatric and Cognitive Difficulties as Indicators of Juvenile Huntington Disease Onset in 29 Patients. *Arch. Neurol.* **2007**, *64*, 813–819. [CrossRef]
9. Bates, G.; Dorsey, R.; Gusella, J.F.; Hayden, M.R.; Kay, C.; Leavitt, B.R.; Nance, M.; Ross, C.A.; Scahill, R.I.; Wetzel, R.; et al. Huntington disease. *Nat. Rev. Dis. Prim.* **2015**, *1*, 15005. [CrossRef]
10. Ruocco, H.H.; Lopes-Cendes, I.; Laurito, T.L.; Min, L.L.; Cendes, F. Clinical presentation of juvenile Huntington disease. *Arq. NeuroPsiquiatr.* **2006**, *64*, 5–9. [CrossRef]
11. Mestre, T.A.; Shannon, K. Huntington disease care: From the past to the present, to the future. *Park. Relat. Disord.* **2017**, *44*, 114–118. [CrossRef] [PubMed]
12. Domaradzki, J. The Impact of Huntington Disease on Family Carers: A Literature Overview. *Psychiatr. Polska* **2015**, *49*, 931–944. [CrossRef]
13. Nance, M.A. Huntington disease: Clinical, genetic, and social aspects. *J. Geriatr. Psychiatry Neurol.* **1998**, *11*, 61–70. [CrossRef] [PubMed]

14. Rohs, G.; Klimek, M.L. Ethical, social and legal issues in Huntington disease: The nurse's role. *Axone* **1996**, *17*, 55–59. [PubMed]

15. Galjaard, H. Toekomstige ontwikkelingen in het erfelijkheidsonderzoek II. Psychologische en maatschappelijke aspecten. *Ned Tijdschr Geneeskd* **1997**, *141*, 2438–2443. (In Dutch)

16. Rawlins, S.M.; Wexler, N.S.; Wexler, A.; Tabrizi, S.J.; Douglas, I.J.; Evans, S.J.; Smeeth, L. The Prevalence of Huntington's Disease. *Neuroepidemiology* **2016**, *46*, 144–153. [CrossRef]

17. Pringsheim, T.; Wiltshire, K.; Day, L.; Dykeman, J.; Steeves, T.; Jette, N. The incidence and prevalence of Huntington's disease: A systematic review and meta-analysis. *Mov. Disord.* **2012**, *27*, 1083–1091. [CrossRef]

18. Evans, S.J.; Douglas, I.; Rawlins, S.M.; Wexler, N.S.; Tabrizi, S.J.; Smeeth, L. Prevalence of adult Huntington's disease in the UK based on diagnoses recorded in general practice records. *J. Neurol. Neurosurg. Psychiatry* **2013**, *84*, 1156–1160. [CrossRef]

19. Ohlmeier, C.; Saum, K.-U.; Galetzka, W.; Beier, D.; Gothe, H. Epidemiology and health care utilization of patients suffering from Huntington's disease in Germany: Real world evidence based on German claims data. *BMC Neurol.* **2019**, *19*, 318. [CrossRef]

20. Hayden, M.R. *Huntington's Chorea*; Springer: London, UK, 1981; ISBN 978-1-4471-1310-2.

21. Van Dijk, J.G.; Van Der Velde, E.A.; Roos, R.A.C.; Bruyn, G.W. Juvenile Huntington disease. *Qual. Life Res.* **1986**, *73*, 235–239. [CrossRef]

22. Sprenger, G.P.; Van Der Zwaan, K.F.; Roos, R.A.; Achterberg, W.P. The prevalence and the burden of pain in patients with Huntington disease. *Pain* **2019**, *160*, 773–783. [CrossRef] [PubMed]

23. Quarrell, O.; O'Donovan, K.L.; Bandmann, O.; Strong, M. The Prevalence of Juvenile Huntington's Disease: A Review of the Literature and Meta-Analysis. *PLoS Curr.* **2012**, *4*, e4f8606b742ef3. [CrossRef] [PubMed]

24. S2k-Leitlinie Chorea/Morbus Huntington. Available online: www.dgn.org/leitlinien (accessed on 29 April 2020).

25. Bonelli, R.M.; Wenning, G.K. Pharmacological management of Huntington's disease: An evidence-based review. *Curr. Pharm. Des.* **2006**, *12*, 2701–2720. [CrossRef] [PubMed]

26. Shannon, K.M.; Fraint, A. Therapeutic advances in Huntington's Disease. *Mov. Disord.* **2015**, *30*, 1539–1546. [CrossRef] [PubMed]

27. Andresen, J.; Gayán, J.; Djoussé, L.; Roberts, S.; Brocklebank, D.; Cherny, S.S.; Cardon, L.R.; Gusella, J.F.; Macdonald, M.E.; Myers, R.; et al. The Relationship Between CAG Repeat Length and Age of Onset Differs for Huntington's Disease Patients with Juvenile Onset or Adult Onset. *Ann. Hum. Genet.* **2006**, *71*, 295–301. [CrossRef]

28. Fusilli, C.; Migliore, S.; Mazza, T.; Consoli, F.; De Luca, A.; Barbagallo, G.; Ciammola, A.; Gatto, E.M.; Cesarini, M.; Etcheverry, J.L.; et al. Biological and clinical manifestations of juvenile Huntington's disease: A retrospective analysis. *Lancet Neurol.* **2018**, *17*, 986–993. [CrossRef]

29. Hart, E.P.; Marinus, J.; Burgunder, J.-M.; Bentivoglio, A.R.; Craufurd, D.; Reilmann, R.; Saft, C.; Roos, R.A. Better global and cognitive functioning in choreatic versus hypokinetic-rigid Huntington's disease. *Mov. Disord.* **2013**, *28*, 1142–1145. [CrossRef]

30. Oosterloo, M.; Bijlsma, E.K.; De Die-Smulders, C.; Roos, R.A. Diagnosing Juvenile Huntington's Disease: An Explorative Study among Caregivers of Affected Children. *Brain Sci.* **2020**, *10*, 155. [CrossRef]

31. Saft, C.; Lauter, T.; Kraus, P.H.; Przuntek, H.; E Andrich, J. Dose-dependent improvement of myoclonic hyperkinesia due to Valproic acid in eight Huntington's Disease patients: A case series. *BMC Neurol.* **2006**, *6*, 11. [CrossRef]

32. Saft, C.; Von Hein, S.M.; Lucke, T.; Thiels, C.; Peball, M.; Djamshidian, A.; Heim, B.; Seppi, K. Cannabinoids for Treatment of Dystonia in Huntington's Disease. *J. Huntingt. Dis.* **2018**, *7*, 167–173. [CrossRef]

33. Wojtecki, L.; Groiss, S.J.; Ferrea, S.; Elben, S.; Hartmann, C.J.; Dunnett, S.B.; Ross, C.A.; Saft, C.; Südmeyer, M.; Ohmann, C.; et al. A Prospective Pilot Trial for Pallidal Deep Brain Stimulation in Huntington's Disease. *Front. Neurol.* **2015**, *6*, 180. [CrossRef] [PubMed]

34. Connolly, B.S.; E Lang, A. Pharmacological Treatment of Parkinson Disease. *JAMA* **2014**, *311*, 1670–1683. [CrossRef]

35. Gambardella, A.; Muglia, M.; Labate, A.; Magariello, A.; Gabriele, A.L.; Mazzei, R.; Pirritano, D.; Conforti, F.; Patitucci, A.; Valentino, P.; et al. Juvenile Huntington's disease presenting as progressive myoclonic epilepsy. *Neurology* **2001**, *57*, 708–711. [CrossRef] [PubMed]

36. Cloud, L.J.; Rosenblatt, A.; Margolis, R.L.; Ross, C.A.; Pillai, J.A.; Corey-Bloom, J.; Tully, H.M.; Bird, T.; Panegyres, P.K.; Nichter, C.A.; et al. Seizures in juvenile Huntington's disease: Frequency and characterization in a multicenter cohort. *Mov. Disord.* **2012**, *27*, 1797–1800. [CrossRef] [PubMed]

37. Hoffmann, R.; Schröder, N.; Brüne, M.; Von Hein, S.; Saft, C. Obsessive-Compulsive Symptoms are Less Common in Huntington's Disease than Reported Earlier. *J. Huntingt. Dis.* **2019**, *8*, 493–500. [CrossRef] [PubMed]

38. Hässler, F.; Dück, A.; Jung, M.; Reis, O. Treatment of Aggressive Behavior Problems in Boys with Intellectual Disabilities Using Zuclopenthixol. *J. Child Adolesc. Psychopharmacol.* **2014**, *24*, 579–581. [CrossRef]

Article

Diagnosing Juvenile Huntington's Disease: An Explorative Study among Caregivers of Affected Children

Mayke Oosterloo [1,2,*], **Emilia K. Bijlsma** [3], **Christine de Die-Smulders** [4,5] **and Raymund A. C. Roos** [2]

1   Department of Neurology, Maastricht University Medical Center, 6202 AZ Maastricht, The Netherlands
2   Department of Neurology, Leiden University Medical Center, 2333 ZA Leiden, The Netherlands; R.A.C.Roos@lumc.nl
3   Department of Clinical Genetics, Leiden University Medical Center, 2333 ZA Leiden, The Netherlands; E.K.Bijlsma@lumc.nl
4   Department of Clinical Genetics, Maastricht University Medical Center, 6202 AZ Maastricht, The Netherlands; c.dedie@mumc.nl
5   GROW Research Institute for Oncology and Developmental Biology, Maastricht University, 6200 MD Maastricht, The Netherlands
*   Correspondence: mayke.oosterloo@mumc.nl; Tel.: +31-43-387-50-62; Fax: +31-43-387-70-55

Received: 6 February 2020; Accepted: 6 March 2020; Published: 7 March 2020

Abstract: *Objective:* To investigate the reasons for the diagnostic delay of juvenile Huntington's disease patients in the Netherlands. *Methods:* This study uses interpretative phenomenological analysis. Eligible participants were parents and caregivers of juvenile Huntington's disease patients. *Results:* Eight parents were interviewed, who consulted up to four health care professionals. The diagnostic process lasted three to ten years. Parents believe that careful listening and follow-up would have improved the diagnostic process. Although they believe an earlier diagnosis would have benefited their child's wellbeing, they felt they would not have been able to cope with more grief at that time. *Conclusion:* The delay in diagnosis is caused by the lack of knowledge among health care professionals on the one hand, and the resistance of the parent on the other. For professionals, the advice is to personalize their advice in which a conscious doctor's delay is acceptable or even useful.

Keywords: juvenile Huntington's disease; pediatric Huntington's disease; early-onset Huntington's disease; personal experiences; caregivers

---

1. Introduction

Huntington's disease (HD) is an autosomal dominant neurodegenerative disease characterized by unwanted movements, psychiatric disorders, and cognitive deterioration. HD results from an unstable and expanded Cytosine Adenine Guanine (CAG) trinucleotide repeat in the Huntingtin (*HTT)* gene on chromosome 4 [1]. A CAG repeat size of 36 or more is invariably associated with HD. Most patients develop symptoms and signs in adulthood, with a mean onset of 40 years of age.

The juvenile form of Huntington's disease (JHD) is rare. It is defined as HD with an onset of <21 years of age [2]. The Juvenile HD Working Group of the European Huntington's Disease Network (EHDN) recently redefined JHD as pediatric HD with an age of onset ≤18 years [3]. However, as our study was conducted before this redefinition, we still use the old definition. JHD contributes about 5.4% of all HD cases, with the percentage ranging from 1%–15% in several series [4,5]. There is an inverse correlation between the length of the CAG repeat and the age of onset. The longer the CAG repeat, the earlier the disease-associated symptoms start [6,7].

The clinical presentation in children differs from that in adults. The most prevalent clinical features at the presentation of JHD are cognitive impairment and behavioral changes [8,9]. Common presenting motor features are rigidity, gait disorder, and oral motor dysfunction [2,4,10,11]. Chorea is uncommon in children with HD but becomes manifest in the second decade. Clinical features in the first decade are defined as two or more of the following features: declining school performance, seizures, oral motor dysfunction, rigidity and/or gait disorder in combination with a parent with (pre)symptomatic HD or a family history of HD [4]. The clinical features of individuals in the second decade (adolescent) are less well defined in the literature but are more comparable to the adult manifestation. Disease progression in JHD is likely to be faster compared to normal onset [12]. The variable and non-specific clinical presentation, such as declining school performances and behavioral disturbance, may be confused with disorders such as autism spectrum disorders or attention deficit hyperactivity disorder (ADHD) or with the effects of disrupted social and home environments in HD families. This significantly increases the chance of misdiagnosis and/or diagnostic delay. JHD is rare and probably less well recognized than usual-onset HD (30–50 years) by health professionals without specific knowledge of the disease. One study shows a mean delay of 9 years before diagnosis and another shows delays ranging from 0 to 6 years [9,11].

The psychosocial impact and personal experience of parenting a child with JHD have been described in earlier studies [13–16]. They describe the denial of parents at first, but also the awareness something is wrong with their child. Furthermore, these studies highlight the positive and negative experiences of the parents regarding the support they receive from family, friends, and professional caregivers.

Our aim is to investigate the diagnostic process of JHD patients in the Netherlands. We focus on the diagnostic timeline, the experiences of the parents or caregivers during the diagnostic process, and the role of the different health care providers. In this way, we hope to gain insight into the reason for the probable diagnostic delay and how to improve the diagnostic path.

2. Methods

This study employed in-depth semi-structured interviews and interpretative phenomenological analysis (IPA), a well-established experiential approach in health and clinical psychology [17]. IPA's focus on the in-depth examination of the psychological process and descriptions of how individuals deal with life-transforming, or life-threatening events, conditions, or events.

2.1. Participants

Eligible participants were parents and/or caregivers of (living or already deceased) JHD patients in the Netherlands. Participants were recruited through a call published in the magazine of the Dutch Huntington's disease association and through HD specialists. The patients themselves were not interviewed.

In this study, JHD was defined as the onset of symptoms and signs of HD before the age of 21 years [4]. Confirmation of the clinical diagnosis by molecular testing was not necessary for inclusion. If the patient was older than 20 years of age at the time of the interview, the caregivers could still participate.

All subjects gave their informed consent for inclusion before they participated in the study. The study was conducted in accordance with the Declaration of Helsinki, and the protocol was approved by the Ethics Committee of Leiden University Medical Center, The Netherlands (P17.025). Participants could also consent to a semi-structured interview with their child's general physician (GP). Information retrieved from the GP was used to complete the timeline of the diagnostic process.

2.2. Data Collection

The interviews took place at the participants' homes or the out-patient clinic and were conducted by the first author and audio-recorded. All interviews started with the opportunity for participants to ask questions about the aim of the study. The interviews were semi-structured, covering general issues regarding JHD (age of onset, first symptoms), the recognition of JHD symptoms by parents and health professionals, the number of consulted health professionals, and their expertise in HD. Furthermore, questions about the quality and quantity of support from health care professionals and the process of diagnosis were included. The following two questions, directly referring to the process of diagnosis, were always asked:

Which elements of the diagnostic process should be improved?

Should the diagnosis of JHD have been made earlier, and if so, would you and your son/daughter have benefited from it?

2.3. Analysis

Consistent with standard qualitative research techniques, the interviews were based on a topic list, which evolved as the interviews progressed through an iterative process to ensure the questions captured all relevant emerging themes.

The interviews were transcribed ad verbatim and analyzed with IPA. First, several close readings of the transcripts were made, and points of interest and significance noted and coded. Second, these initial comments were used to document emerging themes, aiming at capturing the essential quality of the participants' experiences. Third, a list of themes was compiled and connections were made between the themes. The connected themes were grouped and labeled (superordinate theme). The superordinate themes for each interview were then compared, producing a table of comparative themes.

3. Results

Eight parents/caregivers gave informed consent for an interview. The interviews lasted approximately 60 min each. Two parents contacted the researcher after reading the announcement in the magazine of the Huntington's disease association. Five parents were approached by the treating HD specialist, and one parent was informed by another parent within the JHD community and contacted the researcher themselves. Six individual parents/caregivers and two couples were interviewed. Four of them also gave informed consent for an interview with their child's GP. All identifying information has been changed to protect the privacy of the participating families. Table 1 presents the clinical characteristics of the children. The cultural backgrounds of the parents were alike. Seven participants had one affected child. One had two affected children. The mean age at diagnosis was 16 years, with a range of 9 to 24 (Figure 1). The mean repeat size was 62, with a range from 49 to 83. Individual repeat sizes cannot be provided since parents gave no informed consent for this. In six cases the father was the affected parent and, in three cases, the mother. All but one of the affected parents were deceased at the time of the interview. Six of the nine children were still alive. Six children had or still received HD expert care, three from an HD expert neurologist and three in an HD nursing home (Table 2).

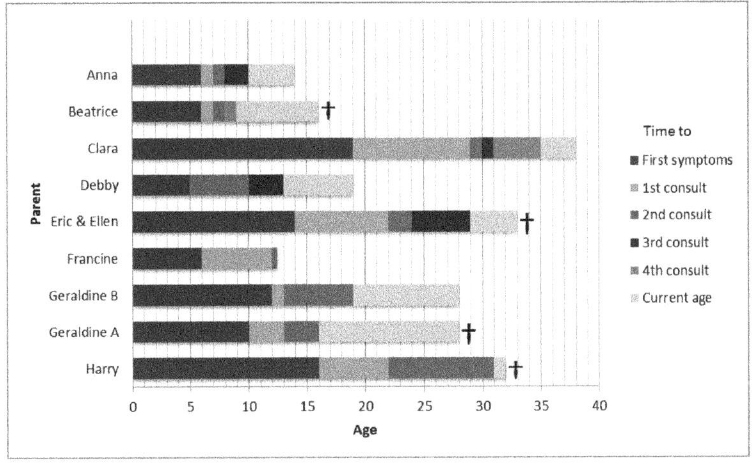

Figure 1. Diagnostic process timeline. † = death.

Table 1. Clinical characteristics of juvenile Huntington's disease (JHD) patients.

Parents	Affected Child	Affected Parent	Age of First Symptoms (y)	Present age (y)	Age of Diagnosis (y)	DNA	MDC	HD Expert Care	Diagnosis
Anna	daughter	F[†]	6–7	14	10	yes	yes	Yes	HD expert neurologist
Beatrice	Son	F	6	16[†]	9	no	yes	Yes	HD expert Neurologist
Clara	Son	F[†]	12–19	38	28	yes	no	Yes	clinical genetics
Debby	Son	F[†]	5	18	13	yes	yes	No	child neurologist
Eric & Ellen	Son	M[†]	14–15	33 [†*]	24	yes	no	Yes	clinical genetics
Francine	stepdaughter	M[†]	6	15	10	yes	yes	Yes	HD expert Neurologist
Geraldine A	daughter	F[†]	10	28 [†*]	16	yes	yes	Yes	MD neurologist
Geraldine B	Son	F[†]	12	28	19	yes	no	No	MD neurologist
Harry	Son	M[†]	16	31[†]	22	yes	yes	No	clinical genetics

F = father; M = mother; † death; * euthanasia; MDC = multidisciplinary care; HD = Huntington's disease; MD = movement disorders.

Table 2. Diagnostic process by consult.

Parent	1st Consult	2nd Consult	3rd Consult	4th Consult
Anna	Psychologist neurological disorder	Psychologist ASD	HD expert HD	
Beatrice	physiotherapist/school doctor neurological disorder	MHS ASD	neurologist/pediatrician spasticity	HD expert HD
Clara	clinical genetics HD	MHS	mental hospital	HD nursing home
Debby	psychologist ADHD	autism expert team ASD	Neurologist HD	
Eric	clinical genetics HD	neurologist	HD nursing home	
Francine	MHS Unknown	HD expert HD		
Geraldine A	Neurologist HD	HD expert HD		
Geraldine B	HD expert ADD	clinical genetics/neurologist HD		
Harry	clinical genetics unknown	assisted living	HD nursing home HD	

All referrals took place after a visit to the general practitioner. All HD experts are neurologists. Abbrevations: HD = Huntington's disease; MHS = mental health services; ASD = Autism Spectrum Disorder

Six superordinate themes emerged from the analysis. They describe the parents' perceptions of the problems associated with the JHD diagnostic process. The six themes are (1) awareness ("something is wrong"), (2) the role of the different health care providers during the diagnostic process, (3) experiences after the diagnosis, (4) the need for support, (5) which elements of the diagnostic process should be improved, and (6) what if the diagnosis had been made earlier? These themes will be highlighted below by means of quotes from the interviewed parents/caregivers.

3.1. Awareness: "Something Is Wrong"

All parents noticed something was wrong with their child when, in retrospect, the first symptoms of the disease presented. The first symptoms varied from decreasing school performance, behavioral or motor symptoms, and sometimes a combination of these. Some initially thought the situation at home with the affected parent or the recent decease of a parent from HD was the cause of the behavior problems.

"My husband was still alive, but he was ill and I thought it's the situation at home. Because there was so much going on at home. I thought it was the stress because dad is at home and also has all these weird complaints." (Anna)

"By the end of elementary school, his grades were so poor. But at that time I thought it was because of the huge problems at home with my husband." (Clara)

These quotes highlight the disruption of the home environment with an ill parent as well as behavioral problems of this parent being seen as a cause of the behavioral changes in children from HD families. Also, puberty was mentioned as the initial presumed cause of the change in the behavior in older children.

"He was aggressive and angry. I thought it was some kind of macho behavior and wasn't quite sure if it had anything to do with HD. We thought it was puberty. But then I started reading about HD and I knew it was Huntington's." (Ellen)

Ellen soon realized the behavioral changes were more likely to be caused by HD rather than puberty. Just like Francine, she is a stepmother. Their husbands both took a lot longer to realize something was wrong.

"It was the first thing I thought. I thought this isn't right. A normal student has the same grades throughout all their school years. Her father and grandmother thought it would be alright and that she was just emotional because her mother had just died." (Francine)

Just like the other parents, Francine's husband believed the death of his daughter's mother was the cause of emotional disruption rather than the start of JHD.

All parents indicated they thought the child's behavior was likely caused by the illness of the affected parent, regardless if they were still alive or already deceased. However, the stepmothers, Francine and Ellen, were convinced that it was symptoms and signs of HD.

3.2. The Role of the Different Health Care Professionals during the Diagnostic Process

3.2.1. Awareness of the Caregiver

After the initial search for explanations, the parents gradually started realizing the signs were probably associated with HD. None of them formally knew about the juvenile onset of HD at the time. They usually consulted their GP for advice.

"Looking back, I think I knew she had Huntington's disease. I just didn't want to hear it, and when a GP tells you it isn't HD, that's simply what you want to hear. When the institute said it was autism, I thought that's just fine, even though I knew it wasn't. It's a very strange way to fool yourself, but you do it anyway." (Anna)

"We went to see our GP, who told us to buy some good shoes, start some therapies. So there was about 2 to 3 years between our first visit to the GP and the neurologist." (Geraldine)

When the thought of HD first crossed their minds, the parents felt anxious and unsure of what to do. They relied on the judgment of the health caregiver. Although most of the parents believed the physician misjudged the situation, they also admit they ignored their own suspicions that it could be JHD.

3.2.2. Positive Support

The parents did experience helpful support from health care professionals in their search for an answer to their child's problems. They mentioned health care professionals who took the problems seriously and/or who offered help and support, even though there was no diagnosis or a diagnosis other than JHD.

"Even though they didn't make a diagnosis, that's different than the feeling of being supported. Sometimes doctors simply don't know what a patient is suffering from, but they do everything in their power to support you, psychologically as well as in other areas." (Anna)

"We started therapy for autism. I asked one of the physiotherapists if he knew what Huntington's disease was. He said he did, and he conducted several tests and put them on video to show a neurologist. After the summer holiday, he came back and said I was right, it was Huntington's disease." (Debby)

3.2.3. Failure of Support

Three children were first diagnosed as having autism or ADHD by several health care professionals, ranging from neurologists to psychologists.

"I remember that psychologist … Eventually, they had to make a diagnosis. She sat there with that green book on her lap, turning the pages back and forth. Finally, she said, well, we'll go with ADHD. I don't believe she was convinced at all." (Debby)

"He started having behavioral problems and we went to see an HD expert neurologist. We already knew he had a 50% chance of HD. But the neurologist ran some tests and said he thought it was a form of Attention Deficit Disorder. He really didn't see any signs of HD. We were told to come back if we started seeing any symptoms of HD." (Geraldine)

The children presented with behavioral changes, such as aggressive behavior, hyperactivity, or obsessive behavior. This led to misdiagnoses such as ADHD and autism. Professional help clearly failed in the period before diagnoses when parents were starting to experience problems with their child. The follow-up period was too short or was lacking entirely. In four cases, the health professionals denied there was anything wrong at all. The parents had a clear need for support from a health care professional on a regular basis.

3.3. Experiences after the Diagnosis

After the diagnosis had been made, most parents and their children received the support they needed. Three of the children received or are now receiving follow-up care from a neurologist with HD expertise. These children were 10 years or younger at the onset of HD. These parents are satisfied with the care provided.

Geraldine's two children received follow-up care from a neurologist with expertise in movement disorders, and she was very satisfied with the support they received.

"We went there twice a year. Our son liked him because I liked him (the neurologist with HD expertise). And there was a special team just for him. That was fantastic. They did so much. Incredible, yes." (Beatrice)

Geraldine and Harry received long-term support from a social worker from one of the clinical genetic departments in the Netherlands. They were also very satisfied with the support from the social worker.

"I must say she was fantastic (social worker). She really took care of us. My daughter bonded with her. She guided us after the diagnosis. And she was the only one my son would speak to. She deserves a huge compliment." (Geraldine)

"She gave me advice. She gave me her mobile number. I once called her on a Sunday, and she rang me back within seconds. I could always tell her my story. Sometimes people complain about the care they get, but hats off for the care we got." (Harry)

It is comforting for parents to receive care from a caregiver who has knowledge of the disease. They feel their problems are taken seriously, and the caregiver is able to help with some of their issues. It is crucial for them that the caregiver knows the course of the disease and is able to support them in the long-term.

3.4. The Need for Support

It was clear to Clara, Eric, and Harry that their child suffered from JHD and they did not feel the need for a formal clinical diagnosis. Their past experiences with caregivers regarding HD were either somewhat negative or they felt caregivers were not able to do anything that would contribute to their child's well-being.

"It was so obvious for us that it was HD, we didn't do anything with it. We went to our GP once, and he had to look it up in a book. And even neurologists don't know what to do with it. They say they know what HD is, but they don't." (Ellen and Eric)

"Well, how shall I put this? I don't know if I'm allowed to say this, but there's nothing a neurologist can do for him." (Harry)

It was clear to Clara that something was wrong or that their child probably suffered from HD, but apart from visits to their GP, she did not feel the need for a referral. She assumed diagnosis was made by DNA testing and "as that is not allowed before the age of 18", there was no use in visiting a specialist.

"In that period, I was aware there was something going on, and then I thought, what am I going to do about it? Because we can't get him tested; he needs to be 18 for that. In fact, nobody made the diagnosis, but for us, it was clear." (Clara)

Anna and Debby felt the need for a diagnosis, but no need for extra support from health care professionals when their child became ill. They explained that the problems at home with an ill or deceased spouse were so huge that there was no place for any kind of support at that time. However, in a later stage, they felt somewhat better and reported they would have liked some support.

"I didn't feel the need. My husband had just died, so there was so much else to deal with at that moment. Looking back, I'd just had time to get back on my feet again when the next thing bombarded us." (Anna)

Clara and Geraldine felt the need for support, but their children did not want medical care or HD expert care.

"The problem was she didn't want HD expert care. They offered us a case manager who came a little too late after we'd already figured it out. We arranged 95% of the care ourselves because she wouldn't let us involve anyone else." (Geraldine)

As described before, many HD patients do not feel the need for support from caregivers or expert centers. However, the family can feel abandoned as they generally do feel the need for support. In their opinion, they could have received support or help even if their child did not want it.

3.5. Which Elements of the Diagnostic Process should be Improved?

Most parents said they felt there was no-one to listen to them and take them seriously. They would have also liked the health professionals to be upfront with them. Four parents visited several health care professionals who offered no follow-up visits, which they clearly would have appreciated.

"I wanted to hear it was nonsense, and he (the GP) told me it was. On the one hand, I was glad, but on the other, I think he should have done something with it, he should have kept an eye on it. Asked us to come in for a check-up a few months later to see how things were going." (Anna)

Even after the diagnosis, some parents did not receive the right support until a later stage of the disease when their child had already been admitted to a nursing home. Clara's son did not want support. However, she believes the health care professionals should have provided more assistance and listened to the parents.

"Well, look at when things get out of hand, and there are no care providers there. It's clear they (patients) don't want them there. They don't want others to see what happens. Take the parents apart, ask how they experience things, how they see things. And then afterward decide how to move on." (Clara)

Parents with an adolescent child with JHD often say their child does not want support from caregivers. This makes it difficult for parents to find long-term care and support. Denial of the disease and problems resulting from the disease are a huge problem in patients with HD. It sometimes takes a lot of patience to persuade HD patients they need long-term care.

Debby was offered support, but at that time, just after the diagnosis was made, it was too much for her as well.

"We contacted a social worker from an HD expertise nursing home, but because we had already figured things out for ourselves, she couldn't add much. However, it would have been good to have a case manager. And for you to be offered a case manager several times instead of once. If they offer this in a very hectic period, it might be too much." (Debby)

Again, parents said they would have appreciated follow-up in time, even though it was not initially wanted or needed.

3.6. What if the Diagnosis of JHD Had Been Made Earlier?

Parents whose children received a diagnosis other than JHD felt they would have been able to treat their children differently with an earlier correct diagnosis. They would have paid more attention and would have had more patience with their child if they had known what was going on.

"I think he would have benefited from it. My husband had just died. Things just kept piling up and I paid less attention and certainly had less patience." (Debby)

"When you know something's wrong with your child, you approach it in a different way. I sometimes thought she was just lazy or didn't feel like doing things. I think it would have made a difference in how I approached her. I wouldn't have been so strict." (Anna)

Furthermore, if the diagnosis had been made earlier, proper medication would have been provided in cases of aggression or other behavioral problems.

"I think that if he'd had medication to suppress the aggression. That would have made life easier for him." (Clara)

Francine believes her stepdaughter would not have been so alone. Before the diagnosis, she went to a regular school. However, because of her changing behavior, she did not have many friends and was isolated.

"Maybe she wouldn't have been so alone and isolated from other children. She was so alone, that was incredibly sad. Of course, I don't know how she experienced it." (Francine)

All the interviewed parents believed an earlier diagnosis would have benefited their children. Especially in the different way they, as parents, would have approached their children. However, as mentioned before, some of the parents had difficulty acknowledging their child might have JHD.

"On the one side, it would have been better if the GP had taken the suspicions of the psychotherapist seriously. He could have sent her to a neurologist. That would have been a more logical decision than telling me not to worry. I suppose it's difficult for doctors to say or … it's logical really, in case of such a terrible disease and knowing a child has a 50% chance, you would think they'd do something if there's any kind of neurological issue. On the other side, I don't know if I would have been able to handle it at the time." (Anna)

An earlier diagnosis may not always have contributed to their well-being as parents. They felt they would not have been able to cope with any more grief or problems than they already had at that time.

4. Discussion

Our study describes the experiences of parents who have cared for or are still caring for a child with JHD, in particular, in regard to the diagnostic process of JHD.

The parents described denial and, afterwards, being aware something was wrong with their child. This denial and awareness are clear phenomena mentioned in previous studies on JHD [13–16]. The most commonly perceived reasons for delayed diagnosis reported in an Australian survey among parents of children living with rare diseases was the lack of knowledge among health professionals [18]. In line with this, we found that the unawareness or lack of knowledge of this serious neurological disease among health care professionals led to a delay in diagnosis. Parents believe that careful listening, attention, and clinical follow up would have improved the diagnostic process.

However, we are the first to report that, for some parents, a period of denial was helpful. They indicated that if the diagnosis JHD had been made shortly after the presentation of the first symptoms, this would have been too much for them. In a way, growing slowly towards the realization of their child being affected by JHD was helpful.

The fact that some of them were recently confronted with a sick or recently deceased spouse contributed to the denial. They would not have been able to manage an earlier diagnosis. In fact, the delay gave them the opportunity to process and face the problems with ill family members consecutively. Some parents visited a neurologist with expertise in HD and would have had the opportunity to discuss their suspicions with the neurologist. They chose not to do so.

So, it seems that shortening the diagnostic delay will not be helpful for every parent. For some children, the mixed feelings of their parents had probably added to a delay in diagnosis. The parents believed their child probably would have benefited from an earlier diagnosis as personalized care and pharmaceutical treatment could have been started earlier, and they would have been more patient with their child.

All children with JHD either received or were offered multidisciplinary care after the diagnosis. However, some parents turned down the offer of help because they either had so many problems that any form of support would have been too much, or their past experience with their affected spouse led them to believe this support would not be of much value.

It seems difficult for clinicians with little or no experience with HD to diagnose JHD. However, HD specialists are still faced with the dilemma if non-specific signs symptoms are caused by JHD or a number of other explanations for the behavioral problems. Therefore, informing the HD community about the existence of JHD could help make parents more aware of the possible signs of JHD. This gives parents the opportunity to gain information if they wish to have it. Better information would also improve their understanding of the condition and of what to expect in the future, and it would probably help them better manage the challenges they face [15].

For clinicians specialized in HD, we recommend a careful follow-up if children from HD families present with non-specific symptoms such as decreased school performance, hyperactivity, and/or behavioral changes [4]. Children from HD families who present with such symptoms should be evaluated longitudinally, at least on an annual basis. Furthermore, we recommend being upfront about whether symptoms and signs might be due to JHD or not. On the other hand, we believe it is important to judge whether parents are able to cope with the diagnosis. Also, it is good to be aware that parents appreciate it when help and support are offered more than once, as the need for support can change during the course of follow-up.

We conclude that the delay in diagnosis is caused by the lack of knowledge among health care professionals on the one hand and the resistance of the parent on the other. In our opinion, this is a new and important finding that has not been described before. For health care professionals, diagnosing JHD is walking a tightrope. As for some parents, an earlier diagnosis would be too distressing; it is important to check whether parents are ready for the information. If not, keep in contact and try again at a later appointment. We hope these findings will be helpful for clinicians, caregivers, and the HD community in contributing to the well-being of these children and their parents. For professionals, our advice is to personalize their advice, in which a conscious doctor's delay is acceptable or even useful.

Author Contributions: Conceptualization, M.O., R.A.C.R. and C.d.D.-S.; Methodology, M.O., E.K.B., C.d.D.-S., R.A.C.R.; Formal Analysis, M.O.; Investigation, M.O.; Resources, M.O., R.A.C.R.; Data Curation, M.O.; Writing—Original Draft Preparation, M.O.; Writing—Review and Editing M.O., E.K.B., C.d.D.-S., R.A.C.R.; Validation, M.O., R.A.C.R.; Visualization, M.O.; Supervision, E.K.B., C.d.D.-S., R.A.C.R. All authors have read and agreed to the published version of the manuscript.

Conflicts of Interest: Raymund Roos is an advisor for Unique, and received a grant for clinical trial from TEVA. The other authors declare they have no conflict of interest concerning the research related to this manuscript.

References

1. The Huntington's Disease Collaborative Research Group. A novel gene containing a trinucleotide repeat that is expanded and unstable on Huntington's disease chromosomes. *Cell* **1993**, *72*, 971–983. [CrossRef]
2. Van Dijk, J.G.; van der Velde, E.A.; Roos, R.A.; Bruyn, G.W. Juvenile Huntington disease. *Hum. Genet.* **1986**, *73*, 235–239. [CrossRef] [PubMed]
3. Quarrell, O.W.J.; Nance, M.A.; Nopoulos, P.; Reilmann, R.; Oosterloo, M.; Tabrizi, S.J.; Furby, H.; Saft, C.; Roos, R.A.C.; Squitieri, F.; et al. Defining pediatric huntington disease: Time to abandon the term Juvenile Huntington Disease? *Mov. Disord. Off. J. Mov. Disord. Soc.* **2019**. [CrossRef] [PubMed]
4. Nance, M.A. Genetic testing of children at risk for Huntington's disease. US Huntington Disease Genetic Testing Group. *Neurology* **1997**, *49*, 1048–1053. [CrossRef] [PubMed]
5. Quarrell, O.; O'Donovan, K.L.; Bandmann, O.; Strong, M. The Prevalence of Juvenile Huntington's Disease: A Review of the Literature and Meta-Analysis. *PLoS Curr.* **2012**, *4*. [CrossRef] [PubMed]
6. Andrew, S.E.; Goldberg, Y.P.; Kremer, B.; Telenius, H.; Theilmann, J.; Adam, S.; Starr, E.; Squitieri, F.; Lin, B.; Kalchman, M.A.; et al. The relationship between trinucleotide (CAG) repeat length and clinical features of Huntington's disease. *Nat. Genet.* **1993**, *4*, 398–403. [CrossRef] [PubMed]
7. Telenius, H.; Kremer, H.P.; Theilmann, J.; Andrew, S.E.; Almqvist, E.; Anvret, M.; Greenberg, C.; Greenberg, J.; Lucotte, G.; Squitieri, F.; et al. Molecular analysis of juvenile Huntington disease: The major influence on (CAG)n repeat length is the sex of the affected parent. *Hum. Mol. Genet.* **1993**, *2*, 1535–1540. [CrossRef] [PubMed]
8. Koutsis, G.; Karadima, G.; Kladi, A.; Panas, M. The challenge of juvenile Huntington disease: To test or not to test. *Neurology* **2013**, *80*, 990–996. [CrossRef] [PubMed]
9. Ribai, P.; Nguyen, K.; Hahn-Barma, V.; Gourfinkel-An, I.; Vidailhet, M.; Legout, A.; Dode, C.; Brice, A.; Durr, A. Psychiatric and cognitive difficulties as indicators of juvenile huntington disease onset in 29 patients. *Arch. Neurol.* **2007**, *64*, 813–819. [CrossRef] [PubMed]
10. Nance, M.A.; Myers, R.H. Juvenile onset Huntington's disease–clinical and research perspectives. *Ment. Retard. Dev. Disabil. Res. Rev.* **2001**, *7*, 153–157. [CrossRef] [PubMed]
11. Gonzalez-Alegre, P.; Afifi, A.K. Clinical characteristics of childhood-onset (juvenile) Huntington disease: Report of 12 patients and review of the literature. *J. Child Neurol.* **2006**, *21*, 223–229. [PubMed]
12. Quarrell, O.W.; Nance, M.A.; Nopoulos, P.; Paulsen, J.S.; Smith, J.A.; Squitieri, F. Managing juvenile Huntington's disease. *Neurodegener. Dis. Manag.* **2013**, *3*. [CrossRef]
13. Smith, J.A.; Brewer, H.M.; Eatough, V.; Stanley, C.A.; Glendinning, N.W.; Quarrell, O.W. The personal experience of juvenile Huntington's disease: An interpretative phenomenological analysis of parents' accounts of the primary features of a rare genetic condition. *Clin. Genet.* **2006**, *69*, 486–496. [CrossRef] [PubMed]
14. Brewer, H.M.; Smith, J.A.; Eatough, V.; Stanley, C.A.; Glendinning, N.W.; Quarrell, O.W. Caring for a child with Juvenile Huntington's Disease: Helpful and unhelpful support. *J. Child. Health Care* **2007**, *11*, 40–52. [CrossRef] [PubMed]
15. Brewer, H.M.; Eatough, V.; Smith, J.A.; Stanley, C.A.; Glendinning, N.W.; Quarrell, O.W. The impact of Juvenile Huntington's Disease on the family: The case of a rare childhood condition. *J. Health Psychol.* **2008**, *13*, 5–16. [CrossRef] [PubMed]

16. Eatough, V.; Santini, H.; Eiser, C.; Goller, M.L.; Krysa, W.; de Nicola, A.; Paduanello, M.; Petrollini, M.; Rakowicz, M.; Squitieri, F.; et al. The personal experience of parenting a child with juvenile Huntington's disease: Perceptions across Europe. *Eur. J. Hum. Genet.* **2013**, *21*, 1042–1048. [CrossRef] [PubMed]
17. Smith, J.A. *Qualitive Psychology*; Sage: London, UK, 2003.
18. Zurynski, Y.; Deverell, M.; Dalkeith, T.; Johnson, S.; Christodoulou, J.; Leonard, H.; Elliott, E.J.; APSU Rare Diseases Impacts on Families Study group. Australian children living with rare diseases: Experiences of diagnosis and perceived consequences of diagnostic delays. *Orphanet J. Rare Dis.* **2017**, *12*, 68. [CrossRef] [PubMed]

Review

Therapeutic Advances for Huntington's Disease

Ashok Kumar [1,†], Vijay Kumar [2,*,†], Kritanjali Singh [3], Sukesh Kumar [4], You-Sam Kim [2], Yun-Mi Lee [2] and Jong-Joo Kim [2,*]

[1] Department of Genetics, Sanjay Gandhi Post-Graduate Institute of Medical Sciences, Lucknow 226014, UP, India; chemistry.ashok83@gmail.com
[2] Department of Biotechnology, Yeungnam University, Gyeongsan, Gyeongbuk 38541, Korea; samsam5332@naver.com (Y.-S.K.); ymlee@yu.ac.kr (Y.-M.L.)
[3] Central Research Station, Subharti Medical College, Swami Vivekanand Subharti University, Meerut 250002, India; skritanjali@gmail.com
[4] PG Department of Botany, Nalanda College, Bihar Sharif, Magadh University, Bihar 824234, India; kumarsukesh92@gmail.com
* Correspondence: vijaykumarcbt@ynu.ac.kr (V.K.); kimjj@ynu.ac.kr (J.-J.K.)
† Contributed equally as co-first authors.

Received: 5 November 2019; Accepted: 10 January 2020; Published: 12 January 2020

Abstract: Huntington's disease (HD) is a progressive neurological disease that is inherited in an autosomal fashion. The cause of disease pathology is an expansion of cytosine-adenine-guanine (CAG) repeats within the huntingtin gene (*HTT*) on chromosome 4 (4p16.3), which codes the huntingtin protein (mHTT). The common symptoms of HD include motor and cognitive impairment of psychiatric functions. Patients exhibit a representative phenotype of involuntary movement (chorea) of limbs, impaired cognition, and severe psychiatric disturbances (mood swings, depression, and personality changes). A variety of symptomatic treatments (which target glutamate and dopamine pathways, caspases, inhibition of aggregation, mitochondrial dysfunction, transcriptional dysregulation, and fetal neural transplants, etc.) are available and some are in the pipeline. Advancement in novel therapeutic approaches include targeting the mutant huntingtin (mHTT) protein and the *HTT* gene. New gene editing techniques will reduce the CAG repeats. More appropriate and readily tractable treatment goals, coupled with advances in analytical tools will help to assess the clinical outcomes of HD treatments. This will not only improve the quality of life and life span of HD patients, but it will also provide a beneficial role in other inherited and neurological disorders. In this review, we aim to discuss current therapeutic research approaches and their possible uses for HD.

Keywords: Huntington's disease; CAG repeat; mutant huntingtin (mHTT); therapeutics; neurodegeneration

---

1. Introduction

Huntington's disease (HD) is genetically inherited in an autosomal dominant fashion. It is a fatal neurodegenerative disease, caused by an abnormal triplet repeat expansion of CAG (cytosine-adenine-guanine) within the huntingtin (*HTT*) gene on chromosome 4p16.3, causing a mutated huntingtin protein (mHTT) [1–5]. HD is predominantly characterized by adult-onset, progressive motor dysfunction, cognitive impairment and psychiatric symptoms (depression, anxiety, obsessive-compulsive disorder, and psychosis). Chorea, incoordination, and rigidity are common motor symptoms due to neurotoxicity of mHTT, leading to brain atrophy of the striatum, thalamus, cerebellum, brain stem and cortex [6–9]. Clinically, HD includes juvenile HD (onset less than 21 years, and marked clinical symptoms), and late-onset HD (after the age of 60 years) [10–12]. Alcohol, drug, and tobacco abuse were associated with earlier onset of HD, and hasten motor onset in women. These abuses have more significant associations in females than in males [13,14]. Children with CAG repeats

≥39, had significantly lower measures of head circumference, weight, and body mass index [15–17]. Disrupted sleep, tics, pain, itching, and psychosis are the common symptoms of juvenile HD [18].

Presently, there is no remedy for HD, and the disease progresses manifests with a presumed continuation of 15–20 years after the appearance of the first symptom [12,19]. The identification of novel biomarkers involves the development of new treatment strategies. The current therapy is palliative and does not change the course of the disease. Tetrabenazine (TBZ; Xenazine™) was approved for the remedy of chorea in HD by the U.S. food and drug administration (FDA). Additionally, the deuterated version of TBZ, deutetrabenazine (AUSTEDO™), has an improved pharmacokinetic profile and was recently approved by the FDA for the treatment of Huntington chorea. In the last review [20], we discussed different promising agents in the treatment of HD, and their phases under clinical trial. Here we describe updates related to these promising agents which will cure HD.

2. Pathogenesis of the HD

HD is a monogenic disease with prevalence of about 1 in 7,500 individuals in the general population [21,22]. The normal allele has less than 27 CAG repeats and intermediate alleles have 27–35 repeats. CAG repeats of 36–39 will develop HD with less penetrance. Individuals who have 40 or more CAG repeats will develop HD with full penetrance. It is also reported that the higher the CAG expansion, the earlier the onset and the greater the disease severity [12,23]. Kremer et al. reported the largest expansion of 121 trinucleotides [24]. CAG codon encodes glutamine α-amino acid (symbol Gln or Q). Glutamine ($C_5H_{10}N_2O_3$) is synthesized from glutamate and ammonia by the enzyme glutamine synthetase. It is mainly produced in muscle, the lungs, and the brain and acts as a precursor to the neurotransmitter glutamate [25]. CAG has glutamine amino acids within the *HTT* gene and it is not toxic in itself. However, the polyglutamine expansion involves the formation of aggregate and ultimately becomes toxic. It is the principal factor for the manifestation of HD because aggregates are never a remarkable feature in the brain of normal subjects [26,27]. Aggregate formations are accountable for secondary problems, like inflammatory responses (altered cytokine and nitric oxide level), mitochondrial dysfunction (imbalanced level of free radicals and oxidative stress markers), nuclear cleavage, apoptosis, excitotoxicity, transcriptional altered regulation, and lastly, are responsible for the altered neuropathological feature (cause of cell death/damage) (Figure 1). Approximately 70% of the variation of the disease is due to expanded CAG repeats, while 13% of the variation is due to polymorphisms in the GRIK2 gene [28]. These depict the importance of secondary factors that affect disease onset, its severity, and possible output.

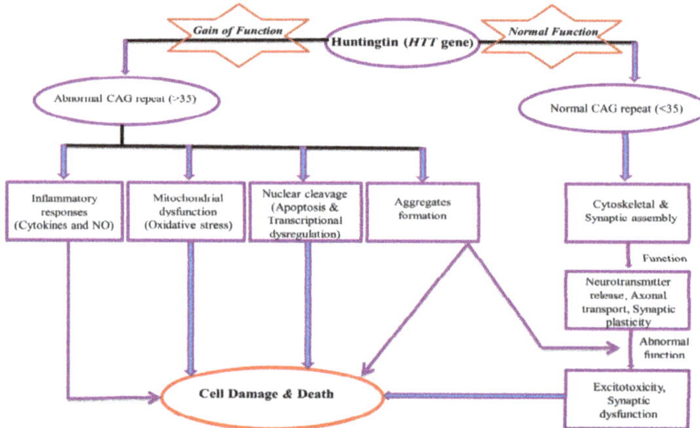

Figure 1. Mechanism of Toxicity of Huntingtin (*HTT*) gene. NO (Nitric Oxide), CAG (cytosine-adenine-guanine).

3. Therapeutic Update

Currently, many drugs are under clinical trial. In the following subsection, we discuss their therapeutic status and their potential role in treatment. These drugs are summarized in Table 1 and Figure 2.

3.1. Drugs against Excitotoxicity

3.1.1. Riluzole and Memantine Drug

Riluzole is a glutamate inhibitor that reduces abnormal movement in amyotrophic lateral sclerosis (ALS) patients [29,30]. In a double-blinded trial, riluzole did not decrease symptoms of HD, nor was it neuroprotective [31].

Memantine is an antagonist of extrasynaptic N-methyl-D-aspartate (NMDA) receptors and is used for the treatment of moderate-severe dementia in Alzheimer's disease (AD). It diminishes striatal cell death, hinders disease progression and improves cognitive function related to HD [32,33]. The combination of memantine and risperidone prevented the expected progression of motor symptoms, cognitive decline, and psychosis over a 6-month study period [34]. However, memantine dosing may be critical, as rodents on low-dose memantine had decreased pathology, while a high-dose of memantine worsened rodent outcomes and possibly promoted cell death [35–37].

3.1.2. Tetrabenazine (TBZ) and Deutetrabenazine

TBZ inhibits the dopamine pathway by inhibiting vesicular monoamine transporter (VMAT) type 2 and consequently decreases available dopamine in the synapse and its interaction with postsynaptic dopamine receptors [38–40]. Deutetrabenazine contains a deuterium atom and is a novel inhibitor of VMAT2. In indirect treatment comparison studies, deutetrabenazine was found to have a favorable tolerability profile compared to tetrabenazine [41]. In mouse models, TBZ ameliorated chorea and other motor symptoms, and reduced striatal neuronal cell loss [38].

3.2. Targeting Caspase Activities and Huntingtin Proteolysis

Minocycline

Minocycline is a tetracycline analog and can cross the blood–brain barrier (BBB) and inhibits the expression of caspase-3 and caspase-1 [42,43]. Treatment with minocycline proved to be neuroprotective, and to improve the disease phenotype [30,42,44]. A human trial study observed motor (unified HD rating scale (UHDRS)), and cognitive (mini-mental state examination (MMSE)) improvement in 14 HD patients who took 100 mg of minocycline for 6 months [43]. This study was continued for another 18 months, and it was found that MMSE, TMS, total functional capacity (TFC) and independence scale were all stabilized after treatment, reducing the expected decline in these measures. There was also a decrease in psychiatric symptoms at 24 months, which was not apparent after 6 months of treatment [44]. In a pilot study, Thomas et al. found improvement in MMSE, UHDRS, and abnormal involuntary movements scale (AIMS), in 30 patients with HD who were given minocycline for 6 months [45].

3.3. Targeting HTT Aggregation and Clearance

3.3.1. Congo Red and Trehalose

Congo red dye binds preferably to β-sheets with amyloid fibrils. When injected into HD mice, it preserved normal protein synthesis and degradation, and improved motor functions. This dye promotes clearance of expanded polyQ repeats and inhibits polyglutamine oligomer formation through the disruption of preformed oligomers. Congo red dye also prevented ATP depletion and caspase activation [46–48].

Trehalose disaccharide inhibited the formation of nuclear inclusions, improved altered motor function and was associated with a high rate of survival in R6/2 mice without causing harmful side effects [49–51] (Table 1 and Figure 2).

3.3.2. Compound C2–8

Compound C2–8 inhibits polyglutamine aggregates in brain slices and cell cultures. It has improved motor function, decreased the amount of neuronal atrophy, and decreased the size of the mHTT aggregates in R6/2 mice [52,53]. There is no ongoing human trial using this compound currently listed on clinicaltrials.gov.

3.3.3. Rapamycin

mTOR is a protein kinase that phosphorylates many proteins and plays a key role in various cellular functions (like autophagy and transcription). mTOR interacts with mHTT and localizes to these polyglutamine aggregates, and thus sequestration of mTOR reduces the activity of mTOR, resulting in a decrease in autophagy and a decrease in the clearance of mHTT. mTOR phosphorylates S6K1 (a key regulator of cell volume), therefore mHTT-related impairment of mTOR may account for the brain atrophy in HD. Rapamycin (which inhibits mTOR and consequently induces autophagy) decreased mHTT aggregates and improved neuronal survival in the drosophila HD model. Rapamycin also improved motor performance and decreased striatal neuropathology in mouse models of HD [54–56].

Table 1. Recent status of Huntington's disease (HD) drug therapy.

Drug/Reagent	Primary Target (Mechanism of Action)	Status and Principal Result	Ref.
(1) Drugs against excitotoxicity			
Riluzole	Glutamate release inhibitor	Does not show efficacy in human trails	[31]
Memantine	N-methyl-D-aspartate (NMDA) receptor inhibitor	Demonstrated efficacy in human trial	[32,33]
Tetrabenazine (TBZ)	Dopamine pathway (Vesicular monoamine transporter 2 inhibitor)	Approved by food and drug administration (FDA) for treatment of chorea in HD	[38–40]
(2) Targeting Caspase and huntingtin (HTT) proteolysis			
Minocycline	Caspase-dependent and independent neurodegenerative pathway	Inhibits caspase-1 and -3 mRNA upregulation, and decreases inducible nitric oxide synthetase activity	[42,44,45]
(3) Targeting *HTT* aggregation and clearance			
Congo red and Trehalose	Aggregation	Showed efficacy in a rodent model	[46,49]
Compound C2–8	Aggregation	Showed efficacy in a rodent model	[53]
Rapamycin	Aggregation mammalian target of rapamycin (mTOR) inhibitor	Showed efficacy in a rodent model	[54,55]
(4) Targeting mitochondrial dysfunction			
Creatine	Mitochondrial dysfunction	Attained futility in human trial	[57]
CoQ10	Mitochondrial dysfunction	Attained futility in human trial	[58]
Eicosapentaenoic acid (EPA)	Mitochondria dysfunction	A mixed scenario of positive and negative trial	[59]
Cystamine and mitochondrial permeability transition pore blockers	Mitochondrial dysfunction	Showed efficacy in a rodent model	[60]
Meclizine drug	Mitochondrial dysfunction	Showed efficacy in the fly model	[61]

Table 1. *Cont.*

Drug/Reagent	Primary Target (Mechanism of Action)	Status and Principal Result	Ref.
(5) Targeting transcriptional dysregulation			
Sodium phenylbutyrate	Transcriptional deregulation histone deacetylase inhibitor	Showed efficacy in a rodent model	[62]
HDACi4b (a pimelic diphenylamide HDAC inhibitor)	Transcriptional deregulation histone deacetylase inhibitor	Showed efficacy in a rodent model	[63]
Suberoylanilide hydroxamic acid	Transcriptional deregulation histone deacetylase inhibitor	Showed efficacy in a rodent model	[64]
Mithramycin and chromomycin	Transcriptional deregulation G-C-rich DNA binding antibiotic	Showed efficacy in a rodent model	[65]
(6) Targetting mutant huntingtin (mHTT)			
RNA interference and antisense oligonucleotide (ASO)	Blocks transcription of mHTT	Showed efficacy in a rodent model	[6]
Artificial peptides and intrabodies	Targeting proline-rich domain of *HTT*	Showed efficacy in a rodent model	[66]
(7) Therapies targeting nucleic acid			
Zinc finger protein	Reduced mutant protein expression	Showed efficacy in a rodent model	[67]
CRISPR-Cas9	Excision of CAG repeat and, reduction of mutant *HTT*	Showed efficacy in a rodent model	[68,69]
ASO approach (IONIS-HTTRX, Peptide conjugated ASOs)	Reduction in *HTT* mRNA and protein	Showed efficacy in a rodent model	[70,71]
RNAi approach (siRNA, shRNA, and miRNA)	Improves motor and neuropathological abnormalities, silencing of *HTT*	Showed efficacy in a rodent model	[6,72]
Novel Viral Vectors (AAV1, AAV5, AAV9, AAV-PHP.B, CREATE)	Widespread transduction of cells	Showed efficacy in primate and rodent models	[73,74]
(8) Other therapeutics			
Ubiquilin	Reduces mHTT aggregation	Showed efficacy in a rodent model	[75,76]
miRNA	Silence *HTT*	Testing in rodent and nonhuman primates	[6,72]
Chaperonins	Decrease mHTT aggregation	Showed efficacy in a rodent model	[77,78]
AFQ056	The antagonist for glutamate receptor 5	Showed no improvement in chorea in a clinical trial	[79]
BN82451	Decreases glutamate release by blocking Na+ channels	Showed efficacy in a rodent model	[80,81]
Antipsychotic drug	Block or modulate dopamine receptors	Under phase III trial	-
Antiapoptotic drug	Cleave mHTT	Effective in mice model	[82,83]
Diet	Delay onset of disease	Effective result but requires further evaluation	[84]

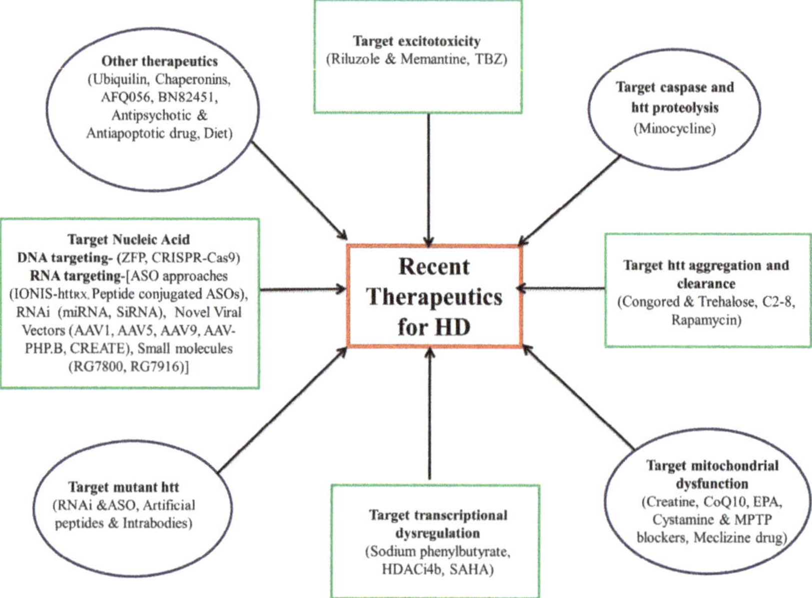

Figure 2. Recent advancement in the therapeutics for Huntington's disease. TBZ (Tetrabenazine); EPA (eicosapentaenoic acid); MPTP (1-methyl-4-phenyl-1,2,3,6-tetrahydropyridine); SAHA (suberoylanilide hydroxamic acid); HDACi4b (histone deacetylase inhibitors); RNAi (RNA interference); ASO (antisense oligonucleotide); ZFP (zinc finger protein); CRISPR-Cas9 (clustered regularly interspaced short palindromic repeats); miRNA (micro RNA); siRNA (small interfering RNA).

3.4. Targeting Mitochondrial Dysfunction

3.4.1. Creatine

Creatine (with antioxidant properties) reduced serum levels of 8-hydroxy-2′-deoxyguanosine (8-OH-2′-dG) in HD patients [85,86] and is safe and tolerable at a dose of 15 g twice daily [87]. In a trial study (with HD patients), receiving a dose of 8 g/day of creatine was secure and well-tolerated but produced no marked change on the UHDRS scale [86]. A randomized double-blind study trial measuring TFC (up to 40 g daily) was terminated early due to futility criteria being reached. The use of creatine fails to delay functional decline in an early manifestation of HD [57]. In another controlled study, creatine (5 g/day; 1 year) treatment resulted in better muscle function capacity in patients with neuromuscular disease but did not show improvements in neuromuscular function and the cognitive status of stage I–III HD patients [88].

3.4.2. Coenzyme Q10

Coenzyme Q10 cofactor is involved in ATP production in the electron transport chain (ETC) of mitochondria, and its supplementation in HD patients may improve mitochondrial function [89]. It was neuroprotective in R6/2 mice, delaying motor deficit, atrophy, and inclusion, and extending survival [90,91]. In a phase III randomized clinical trial, coenzyme Q10 was not effective and the trial was stopped as the futility criteria were reached (http://hdsa.org/wp-content/uploads/2015/01) [58].

3.4.3. Eicosapentaenoic Acid (EPA)

Ethyl-EPA is a derivative of the n-3 polyunsaturated fatty acid EPA, which binds to the peroxisome proliferator-activated receptor of mitochondria [92]. Ethyl-EPA improves the neuronal function by inhibiting caspase and reducing mitochondrial damage by reducing the activity of the c-Jun N-terminal kinase (JNK) pathway [93,94]. Treatment with ethyl-EPA (2 g/day) showed a stable/improved motor function. However, intent-to-treat analysis showed no significant change between ethyl-EPA and placebo for total motor score 4 (TMS–4) subscale in HD patients (stage III) [95]. Patients with fewer CAG repeats showed significant improvement in TMS-4.

In a phase III, double-blind randomized control trial, ethyl-EPA did not show improvement in TMS, cognition or global impression over 6 months. After 6 months, all participants (both those in the treatment and placebo group) were given ethyl-EPA. Those in the original treatment group showed a better motor function (indicated by TMS scores). This suggests that ethyl-EPA needs a longer period before improvement can be observed which might possibly reflect a disease modification [96]. In a recent study, no significant improvement of the treatment group over placebo group was found in measures of TMS or UHDRS scores [59].

3.4.4. Cystamine and MPTP (1-methyl-4-phenyl-1,2,3,6-tetrahydropyridine) Blockers

Both cystamine and MPTP increase the survival effects of HD cells and inhibit oxidative damage [60].

3.4.5. Meclizine

Meclizine, an antihistamine drug, inhibits oxidative metabolism and apoptosis, and is neuroprotective in drosophila model. Energy metabolism deficits and neuronal degeneration are hallmarks of HD, so treatment with meclizine is a potential strategy, especially since it crosses the BBB [20,61] (Table 1 and Figure 2). There is no human clinical trial for this drug listed in clinicaltrials.gov.

3.5. Targeting Transcriptional Dysregulation

3.5.1. Sodium Phenylbutyrate

Administration of sodium phenylbutyrate (an HDAC inhibitor) to N171-82Q symptomatic mice showed less brain atrophy and extended survival rates. Further, it increased and decreased histone acetylation and methylation, respectively, in the rodent brain. It also downregulated caspases involved in apoptosis [62]. A dose-response study highlighted that sodium phenylbutyrate was safe, secure, effective and well-tolerable in HD patients [97].

3.5.2. HDACi4b

HDACi4b (a pimelic diphenylamide HDAC inhibitor) improves motor impairment as well as decreases neurodegeneration in mouse models of HD. Oral administration of HDACi4b to mice showed improvement in these motor deficits. These mice also showed less striatal atrophy and brain-size reduction. HDACi4b reversed hypoacetylation of the H3 histone subunit that occurs in the presence of mHTT, and mRNA expression was returned to normal levels [63].

3.5.3. Suberoylanilide Hydroxamic Acid (SAHA)

Histone acetylation in the brain is increased by SAHA (by inhibiting HDAC) which improves motor impairments in transgenic R6/2 mice. SAHA can be taken orally because it crosses the BBB, however, this has not been tested in humans [64].

3.5.4. Mithramycin and Chromomycin

Treatments with mithramycin and chromomycin (anthracyclin derivatives) promote epigenetic histone modifications in cultured R6/2 and N171-82Q transgenic cell lines, providing a basis for clinical trials for HD [65,98,99].

3.6. Agents Targeting Mutant Huntingtin

3.6.1. RNA Interference (RNAi) and Antisense Oligonucleotide (ASO)

ASO and RNAi execute their knockdown function by allele and nonallele-selective manners [100–102]. As an example, modified ASO (peptide nucleic acid i.e., PNA) enables the selective recognition of the mutant allele and selective inhibition of mHTT expression in human fibroblasts [103]. RNAi reduced neuropathology, improved motor behavior and extended viability in HD [102,104,105].

3.6.2. Intrabodies and Artificial Peptides

In transgenic mouse models of HD like R6/2, N171-82Q, YAC128, and BACHD, treatment with intrabody gene therapy improved body weight, motor function, cognitive, and neuropathological manifestation [66,106].

3.7. Nucleic Acid-Targeting Therapies

3.7.1. Therapies Targeting DNA

Currently, zinc finger proteins (ZFPs) and CRISPR-Cas9 (clustered regularly interspaced short palindromic repeats-CRISPR-associated system) are under investigation.

ZFPs

ZFPs are one of the most abundant protein groups and have various functions, including regulation of DNA, RNA, and protein function. They can bind to specific sequences of DNA and can be used as therapeutic compounds. ZFPs reduce mHTT expression without affecting the expression of other genes/wild-type *HTT* [67,102,107].

CRISPR-Cas9

CRISPR-Cas9 is involved in viral defense mechanisms of bacteria which recognize and destroy foreign DNA. CRISPR-Cas9 is involved in the excision of CAG repeats to make harmless alleles and silence the mHTT expression by insertion of stop codon/missense mutations [68,108–110]. In HD140Q-knockin mice, it was demonstrated that CRISPR-Cas9 can be used to reduce mHTT and improve motor function, but not to increase the lifespan of these mice [69]. Ekman et al. showed that CRISPR-Cas9 can be used for mHTT editing, which can extend survival and improve motor function in the R6 mice following intrastriatal delivery [111].

3.7.2. RNA Targeting Therapies

The four major methods to inhibit the function of mHTT mRNA are: ASOs, RNAi compounds, novel viral vectors, and small-molecule splicing modulators.

ASO Approaches

ASO are single-stranded DNA (ssDNA) molecules that primarily bind to a specific sequence on RNA and regulate post-transcriptional gene expression [112]. The ssDNA diffuses well in the CNS and is taken up by neurons. Therefore, the injection of ASOs into the cerebrospinal fluid (CSF) results in ubiquitous delivery of drugs and suppresses the production of mHTT [70] (Table 1). However, ASO delivery has some side effects, like thrombocytopenia which was observed in some human trials of ASOs [113]. ASO can ameliorate transcriptional dysregulation and reduce the level of mHTT and improve behavior in the YAC128, YAC18, and BACHD mouse models of HD [114,115].

IONIS-HTTRx is an important ASO. It has 12–25 nucleotides and transforms phosphodiester linkages to phosphorothioate. IONIS-HTTRx caused a remarkable reduction in *HTT* mRNA and protein expression [71]. The injection of ASOs (conjugated with peptides), produced wide CNS distribution and longer life span in the spinal muscular atrophy (SMA) mouse model [116]. In a recent study of phase I–IIa trial, HTTRx lessened the concentration of mutant *HTT* in CSF of HD patients. Therefore, ASO compounds not only suppress the expression of *HTT* mRNA and the huntingtin protein in CNS, but also in CSF [117].

RNAi Approaches

RNA interference is a gene-silencing process that uses short interfering RNA (siRNA), short hairpin RNA (shRNA), bi-functional shRNA and microRNA (miRNA). The combination of neural progenitor stem cell therapy and RNAi therapy can ameliorate symptoms in mouse models of HD [118]. In the animal models (R6/1, R62, N171-82Q, RAT AAV-HD70^d) of HD, siRNA, shRNA, and miRNA treatments have been used to reduce neuropathology and improve motor function [6,72,104]. AMT-130 (adeno-associated virus vector) contains an artificial miRNA which produces a huntingtin-lowering molecule. Side effects include peripheral neuropathy observed in clinical trials of siRNA. RNAi has been tested in rodents and its delivery system has been tested in nonhuman primates [119].

Small Molecule Approach

Small molecules like RG7800 showed ocular complications in the Δ7 mouse model of spinal muscular atrophy (SMA) [120], while the phase I trial of the molecule RG7916 (risdiplam) was recently completed (NCT02633709). RG7800 and RG7916 are splicing modifiers, which change the way the pre-mRNA is spliced so that it contains all the information necessary to make a functional protein. They promote the production of a full-length and functional protein from the gene. RG7800 increases the survival motor neuron (SMN) protein level by modifying the splicing of the SMN2 mRNA. RG7800 is shown to promote the inclusion of exon 7 in SMN2 mRNA, generating full-length mRNA, using fibroblasts from an SMA type I patient. In the SMA mouse model, the treatment of RG7800 showed a dose-dependent increase in SMN protein levels [121]. Oral administration of RG7800 in SMA patients increased the functional SMN protein level up to two-fold from baseline [122]. New work is now underway to identify these molecules and their possible role in the lowering of mutant *HTT* gene and protein expression [102,123].

3.8. Other Therapeutics Advancements

3.8.1. Ubiquilin

Ubiquilin overexpression in R6/2 mice decreased aggregation in the hippocampus and cortex and increased lifespan. However, its overexpression did not improve motor symptoms, and did not change the amount of aggregates in the striatum [75,76].

3.8.2. Chaperonins

TRiC (CCT1-CCT8 subunit) is an example of chaperonin which is involved in the folding of about 9–15% of the normal proteins [77] and inhibits aggregation of mHTT [78]. It reduced the number of inclusions, fibrillar oligomers, and insoluble mHTT fragments.

3.8.3. AFQ056

AFQ056 did not improve chorea in a randomized double-blind clinical trial [79].

3.8.4. BN82451

BN82451 inhibits cyclooxygenases and provides antioxidant, anti-inflammatory and neuroprotective effects [80]. It also improved motor function and survival, decreased brain atrophy, neuronal atrophy, and neuronal mHTT inclusions in the R6/2 mice [81]. Recently, a phase II clinical trial has been completed in male HD patients. As per clinicaltrials.gov (NCT02231580), no results have been published.

3.8.5. Antipsychotic Drugs

Antipsychotic drugs are used to treat chorea associated symptoms because they block or modulate dopamine receptors. Many antipsychotic drugs (especially typical antipsychotics) produce motor dysfunction resembling Parkinson's disease (PD). Currently, a phase III trial comparing TBZ with olanzapine and tiapridal is under evaluation [39].

3.8.6. Antiapoptotic Drugs

Caspase cleavage (mainly caspase-3 and -6) occur in mHTT [83]. Mutating the caspase cleavage sites on mHTT leads to neuroprotection and prevents neurodegeneration in yeast artificial chromosome (YAC) mice that express mHTT. Caspase-3 and -6 resistant mice did not develop HD neurodegeneration, indicating that cleavage at these caspase sites plays an important role in neurodegeneration of HD [82,123,124].

3.8.7. Diet

Various studies indicate that a Mediterranean-type diet may delay the onset of other neurodegenerative diseases, like AD, PD, dementia and cognitive impairment [125]. Recently a study highlighted that a Mediterranean-type diet affects the time to HD phenoconversion. In fact, eating high amounts of dairy products was associated with an increased risk of phenoconversion. This may be due to a lower level of urate, which leads to a faster progression and manifestation of HD. These types of diet-related studies need further investigation [84,107,126]. Intermittent fasting promotes autophagy and cleared the mHTT [127].

3.9. Some Promising Clinical Trials

3.9.1. Cysteamine (CYST)

Cysteamine controls oxidation levels through increasing concentration of glutathione, activation of protein catabolism through the hindrance of transglutaminase and induction of heat shock proteins (HSPs) effects [128]. Inhibition of transglutaminase is the putative mode of action and is observed in R6/2 and zQ175 mouse models of HD [129]. The Cysteamine-HD phase II/III trials indicated a delay in the release of cysteamine in HD patients.

3.9.2. Pridopidine

Pridopidine is a modulator of the dopamine 2 receptor [130] and activates the sigma-1 receptor [131]. In the most recent trials of pridopidine, no improvement in motor symptoms (unified Huntington's Disease rating scale - total motor score (UHDRS-TMS)) was observed with placebo [132–134]. High level of pridopidine is found in brain-derived neurotrophic factor (BDNF) and diminishes mHTT aggregate size and improved motor performance in R6/2 mice [135,136]. In an early-stage HD, improvement in the total motor score (TMS) was observed in treated patients [136,137].

3.9.3. Triheptanoin

Triheptanoin is a triglyceride that reverses the metabolic defects in HD by supplying substrates to the Krebs cycle [138]. Recently, a phase II study using triheptanoin was conducted in the early phase of HD patients (listed at clincialtrials.gov under #NCT02453061).

3.9.4. Latrepirdine (Dimebon)

Latrepirdine stabilize and enhance mitochondrial membranes and functions. In a short duration trial, latrepirdine promoted cognitive improvement (MMSE) in mild to moderate HD patients [11,139]. At the time of writing this, a total of 13 clinical trials using latrepirdine have been reported on clinicaltrials. gov (NCT00497159, NCT01085266, NCT00920946, NCT00387270, NCT00988624, NCT00827034, NCT00990613, NCT00824590, NCT00931073, NCT00831506, NCT00788047, NCT00825084).

3.9.5. Amantadine

Amantadine is a weak NMDA receptor blocker [140] and increases dopamine release [141]. Amantadine reduces dyskinesias in HD, without provoking parkinsonism [142].

3.9.6. Lamotrigine

Lamotrigine is an antiepileptic drug and decreases glutamate release by blocking voltage-gated sodium channels [143,144]. It reduces motor symptoms and elevates mood in HD patients [145].

3.9.7. Selisistat

Selisistat is a selective SirT1 inhibitor which removes acetyl groups on proteins, including mHTT. Therefore, blocking the deacetylation of mHTT should activate clearance. In early-stage HD patients, selisistat improved TMS, but not in most measures of cognition, mood, and functionality [146,147].

3.9.8. Tauroursodeoxycholic Acid/Ursodiol

Tauroursodeoxycholic acid (TUDCA) is a bile acid and has antiapoptotic properties in a mouse model of HD. Mice given TUDCA showed less striatal atrophy, apoptosis, and reduced locomotor and sensory-motor defects [148]. A commercially available uroursodeoxycholic acid precursor, ursodiol, has been examined in a phase I trial, but to date, not reported.

3.9.9. Laquinimod

Laquinimod reduces the expression of Bax, responsible for the release of cytochrome C from mitochondria and activation of caspases, causing apoptosis and production of toxic mHTT fragments. This drug improves motor function and reduces depressive behaviors in mice. It is recently undergoing a phase II clinical trial in human HD patients [149]. Laquinimod ameliorates myelination deficiency and behavioral deficits in the YAC128 mouse model of HD [150,151].

3.9.10. Kynurenine Inhibitors

The enzyme indoleamine 2,3 dioxygenase (IDO1) catalyzes the conversion of tryptophan into kynurenine. Kynurenine is then metabolized into 3-hydroxykynurenine (3-HK) and quinolinic acid, both of which are neurotoxic and are increased in HD. In contrast, kynurenine is also metabolized into kynurenic acid, which is neuroprotective. In HD, an imbalance exists between the neurotoxic products and neuroprotective products and targeting the rate-limiting step of IDO1 could effectively shift the balance toward neuroprotective [152]. Kynurenine 3-monooxygenase is the enzyme that catalyzes the conversion of kynurenine into 3-HK. Treating microglial cells from R6/2 mice with a kynurenine 3-monooxygenase inhibitor (Ro 61–8048) showed dramatically reduced 3-HK levels compared to the vector containing cells [153].

4. Conclusions and Future perspectives

The current therapeutic investigations of HD mainly focus on excitotoxicity, dopamine pathway, caspase inhibitors, mHTT aggregation, mitochondrial dysfunction, transcriptional dysregulation, and diet. The application of robust molecular imaging and digital biomarkers may provide a valuable therapeutic boost to the design of clinical trials. Additionally, the increased openness of regulatory agencies for effectiveness will also promote the development of clinical trials. The advancement of modern technologies, and the availability of various promising agents/molecules enable the development of therapies which will further improve the quality of research and outcomes in HD patients. The most promising drugs are those that target the production of mHTT protein and block its actions.

Author Contributions: Conceptualization, A.K., V.K. J.-J.K.; methodology, A.K., V.K.; writing—original draft preparation, A.K., V.K.; writing—review and editing J.-J.K., S.K., V.K., Y.-M.L., Y.-S.K., A.K., K.S. All authors have read and agreed to the published version of the manuscript.

Funding: This research received no external funding.

Conflicts of Interest: The authors declare no conflicts of interest.

Abbreviations

Abnormal involuntary movements scale	AIMS
Alzheimer's disease	AD
Antisense Oligonucleotide	ASO
Blood-brain barrier	BBB
Brain-derived neurotrophic factor	BDNF
Clustered regularly interspaced short palindromic repeats-CRISPR-associated system	CRISPR-Cas9
Cerebrospinal fluid	CSF
Cytosine-adenine-guanine	CAG
Eicosapentaenoic acid	EPA
Electron transport chain	ETC
Food and Drug Administration	FDA
Huntington disease	HD
Huntingtin gene	*HTT*
micro RNA	miRNA
Mini-mental state examination	MMSE
Mutant huntingtin protein	mHTT
Mammalian target of rapamycin	mTOR
1-methyl-4-phenyl-1,2,3,6-tetrahydropyridine	MPTP
N-methyl-D-aspartate	NMDA
Parkinson's disease	PD
RNA Interference	RNAi
small interfering RNA	siRNA
Spinal muscular atrophy	SMA
Suberoylanilide hydroxamic acid	SAHA
Tauroursodeoxycholic acid	TUDCA
Tetrabenazine	TBZ
Total functional capacity	TFC
Total motor score	TMS
Unified HD rating scale	UHDRS
Vesicular monoamine transporter 2	VMAT2
Zinc finger proteins	ZFPs

References

1. Kim, S.D.; Fung, V.S. An update on Huntington's disease: From the gene to the clinic. *Curr. Opin. Neurol.* **2014**, *27*, 477–483. [CrossRef] [PubMed]
2. Ross, C.A.; Aylward, E.H.; Wild, E.J.; Langbehn, D.R.; Long, J.D.; Warner, J.H.; Scahill, R.I.; Leavitt, B.R.; Stout, J.C.; Paulsen, J.S.; et al. Huntington disease: Natural history, biomarkers and prospects for therapeutics. *Nat. Rev. Neurol.* **2014**, *10*, 204–216. [CrossRef] [PubMed]
3. Lee, J.K.; Conrad, A.; Epping, E.; Mathews, K.; Magnotta, V.; Dawson, J.D.; Nopoulos, P. Effect of Trinucleotide Repeats in the Huntington's Gene on Intelligence. *EBioMedicine* **2018**, *31*, 47–53. [CrossRef]

4. Sun, Y.M.; Zhang, Y.B.; Wu, Z.Y. Huntington's Disease: Relationship Between Phenotype and Genotype. *Mol. Neurobiol.* **2017**, *54*, 342–348. [CrossRef] [PubMed]

5. Moss, D.J.H.; Pardinas, A.F.; Langbehn, D.; Lo, K.; Leavitt, B.R.; Roos, R.; Durr, A.; Mead, S.; TRACK-HD investigators; REGISTRY investigators; et al. Identification of genetic variants associated with Huntington's disease progression: A genome-wide association study. *Lancet Neurol.* **2017**, *16*, 701–711. [CrossRef]

6. Harper, S.Q.; Staber, P.D.; He, X.; Eliason, S.L.; Martins, I.H.; Mao, Q.; Yang, L.; Kotin, R.M.; Paulson, H.L.; Davidson, B.L.; et al. RNA interference improves motor and neuropathological abnormalities in a Huntington's disease mouse model. *Proc. Natl. Acad. Sci. USA* **2005**, *102*, 5820–5825. [CrossRef] [PubMed]

7. Hassel, B.; Tessler, S.; Faull, R.L.; Emson, P.C. Glutamate uptake is reduced in prefrontal cortex in Huntington's disease. *Neurochem. Res.* **2008**, *33*, 232–237. [CrossRef]

8. Labbadia, J.; Morimoto, R.I. Huntington's disease: Underlying molecular mechanisms and emerging concepts. *Trends Biochem. Sci.* **2013**, *38*, 378–385. [CrossRef]

9. Chao, T.K.; Hu, J.; Pringsheim, T. Risk factors for the onset and progression of Huntington disease. *Neurotoxicology* **2017**, *61*, 79–99. [CrossRef]

10. Fusilli, C.; Migliore, S.; Mazza, T.; Consoli, F.; De Luca, A.; Barbagallo, G.; Ciammola, A.; Gatto, E.M.; Cesarini, M.; Etcheverry, J.L.; et al. Biological and clinical manifestations of juvenile Huntington's disease: A retrospective analysis. *Lancet Neurol.* **2018**, *17*, 986–993. [CrossRef]

11. Horizon Investigators Of The Huntington Study Group; European Huntington's Disease Network. A randomized, double-blind, placebo-controlled study of latrepirdine in patients with mild to moderate Huntington disease. *JAMA Neurol.* **2013**, *70*, 25–33. [CrossRef] [PubMed]

12. McColgan, P.; Tabrizi, S.J. Huntington's disease: A clinical review. *Eur. J. Neurol.* **2018**, *25*, 24–34. [CrossRef] [PubMed]

13. Byars, J.A.; Beglinger, L.J.; Moser, D.J.; Gonzalez-Alegre, P.; Nopoulos, P. Substance abuse may be a risk factor for earlier onset of Huntington disease. *J. Neurol.* **2012**, *259*, 1824–1831. [CrossRef] [PubMed]

14. Schultz, J.L.; Kamholz, J.A.; Moser, D.J.; Feely, S.M.; Paulsen, J.S.; Nopoulos, P.C. Substance abuse may hasten motor onset of Huntington disease: Evaluating the Enroll-HD database. *Neurology* **2017**, *88*, 909–915. [CrossRef]

15. Lee, J.K.; Mathews, K.; Schlaggar, B.; Perlmutter, J.; Paulsen, J.S.; Epping, E.; Burmeister, L.; Nopoulos, P. Measures of growth in children at risk for Huntington disease. *Neurology* **2012**, *79*, 668–674. [CrossRef]

16. Aylward, E.H.; Liu, D.; Nopoulos, P.C.; Ross, C.A.; Pierson, R.K.; Mills, J.A.; Long, J.D.; Paulsen, J.S.; PREDICT-HD Investigators; Coordinators of the Huntington Study Group. Striatal volume contributes to the prediction of onset of Huntington disease in incident cases. *Biol. Psychiatry* **2012**, *71*, 822–828. [CrossRef]

17. Tereshchenko, A.; Magnotta, V.; Epping, E.; Mathews, K.; Espe-Pfeifer, P.; Martin, E.; Dawson, J.; Duan, W.; Nopoulos, P. Brain structure in juvenile-onset Huntington disease. *Neurology* **2019**, *92*, e1939–e1947. [CrossRef]

18. Moser, A.D.; Epping, E.; Espe-Pfeifer, P.; Martin, E.; Zhorne, L.; Mathews, K.; Nance, M.; Hudgell, D.; Quarrell, O.; Nopoulos, P.; et al. A survey-based study identifies common but unrecognized symptoms in a large series of juvenile Huntington's disease. *Neurodegener. Dis. Manag.* **2017**, *7*, 307–315. [CrossRef]

19. Bates, G.P.; Dorsey, R.; Gusella, J.F.; Hayden, M.R.; Kay, C.; Leavitt, B.R.; Nance, M.; Ross, C.A.; Scahill, R.I.; Wetzel, R.; et al. Huntington disease. *Nat. Rev. Dis. Primers* **2015**, *1*, 15005. [CrossRef]

20. Kumar, A.; Kumar Singh, S.; Kumar, V.; Kumar, D.; Agarwal, S.; Rana, M.K. Huntington's disease: An update of therapeutic strategies. *Gene* **2015**, *556*, 91–97. [CrossRef]

21. Evans, S.J.; Douglas, I.; Rawlins, M.D.; Wexler, N.S.; Tabrizi, S.J.; Smeeth, L. Prevalence of adult Huntington's disease in the UK based on diagnoses recorded in general practice records. *J. Neurol. Neurosurg. Psychiatry* **2013**, *84*, 1156–1160. [CrossRef]

22. Fisher, E.R.; Hayden, M.R. Multisource ascertainment of Huntington disease in Canada: Prevalence and population at risk. *Mov. Disord.* **2014**, *29*, 105–114. [CrossRef] [PubMed]

23. Nance, M.A. Genetics of Huntington disease. *Handb. Clin. Neurol.* **2017**, *144*, 3–14. [CrossRef] [PubMed]

24. Kremer, B.; Goldberg, P.; Andrew, S.E.; Theilmann, J.; Telenius, H.; Zeisler, J.; Squitieri, F.; Lin, B.; Bassett, A.; Almqvist, E.; et al. A worldwide study of the Huntington's disease mutation. The sensitivity and specificity of measuring CAG repeats. *N. Engl. J. Med.* **1994**, *330*, 1401–1406. [CrossRef]

25. Newsholme, P.; Lima, M.M.; Procopio, J.; Pithon-Curi, T.C.; Doi, S.Q.; Bazotte, R.B.; Curi, R. Glutamine and glutamate as vital metabolites. *Braz. J. Med. Biol. Res.* **2003**, *36*, 153–163. [CrossRef] [PubMed]
26. DiFiglia, M.; Sapp, E.; Chase, K.O.; Davies, S.W.; Bates, G.P.; Vonsattel, J.P.; Aronin, N. Aggregation of huntingtin in neuronal intranuclear inclusions and dystrophic neurites in brain. *Science* **1997**, *277*, 1990–1993. [CrossRef]
27. Becher, M.W.; Kotzuk, J.A.; Sharp, A.H.; Davies, S.W.; Bates, G.P.; Price, D.L.; Ross, C.A. Intranuclear neuronal inclusions in Huntington's disease and dentatorubral and pallidoluysian atrophy: Correlation between the density of inclusions and IT15 CAG triplet repeat length. *Neurobiol. Dis.* **1998**, *4*, 387–397. [CrossRef]
28. Lutz, R.E. Trinucleotide repeat disorders. *Semin. Pediatr. Neurol.* **2007**, *14*, 26–33. [CrossRef]
29. Palfi, S.; Riche, D.; Brouillet, E.; Guyot, M.C.; Mary, V.; Wahl, F.; Peschanski, M.; Stutzmann, J.M.; Hantraye, P. Riluzole reduces incidence of abnormal movements but not striatal cell death in a primate model of progressive striatal degeneration. *Exp. Neurol.* **1997**, *146*, 135–141. [CrossRef]
30. Turck, P.; Frizzo, M.E. Riluzole stimulates BDNF release from human platelets. *Biomed. Res. Int.* **2015**, *2015*, 189307. [CrossRef]
31. Landwehrmeyer, G.B.; Dubois, B.; de Yebenes, J.G.; Kremer, B.; Gaus, W.; Kraus, P.H.; Przuntek, H.; Dib, M.; Doble, A.; Fischer, W.; et al. Riluzole in Huntington's disease: A 3-year, randomized controlled study. *Ann. Neurol.* **2007**, *62*, 262–272. [CrossRef] [PubMed]
32. Beister, A.; Kraus, P.; Kuhn, W.; Dose, M.; Weindl, A.; Gerlach, M. The N-methyl-D-aspartate antagonist memantine retards progression of Huntington's disease. In *Focus on Extrapyramidal Dysfunction*; Müller, T., Riederer, P., Eds.; Springer: Vienna, Austria, 2004; pp. 117–122.
33. Lee, S.T.; Chu, K.; Park, J.E.; Kang, L.; Ko, S.Y.; Jung, K.H.; Kim, M. Memantine reduces striatal cell death with decreasing calpain level in 3-nitropropionic model of Huntington's disease. *Brain Res.* **2006**, *1118*, 199–207. [CrossRef] [PubMed]
34. Cankurtaran, E.S.; Ozalp, E.; Soygur, H.; Cakir, A. Clinical experience with risperidone and memantine in the treatment of Huntington's disease. *J. Natl. Med. Assoc.* **2006**, *98*, 1353–1355. [PubMed]
35. Dau, A.; Gladding, C.M.; Sepers, M.D.; Raymond, L.A. Chronic blockade of extrasynaptic NMDA receptors ameliorates synaptic dysfunction and pro-death signaling in Huntington disease transgenic mice. *Neurobiol. Dis.* **2014**, *62*, 533–542. [CrossRef]
36. Okamoto, S.; Pouladi, M.A.; Talantova, M.; Yao, D.; Xia, P.; Ehrnhoefer, D.E.; Zaidi, R.; Clemente, A.; Kaul, M.; Graham, R.K.; et al. Balance between synaptic versus extrasynaptic NMDA receptor activity influences inclusions and neurotoxicity of mutant huntingtin. *Nat. Med.* **2009**, *15*, 1407–1413. [CrossRef]
37. Milnerwood, A.J.; Gladding, C.M.; Pouladi, M.A.; Kaufman, A.M.; Hines, R.M.; Boyd, J.D.; Ko, R.W.; Vasuta, O.C.; Graham, R.K.; Hayden, M.R.; et al. Early increase in extrasynaptic NMDA receptor signaling and expression contributes to phenotype onset in Huntington's disease mice. *Neuron* **2010**, *65*, 178–190. [CrossRef]
38. Wang, H.; Chen, X.; Li, Y.; Tang, T.S.; Bezprozvanny, I. Tetrabenazine is neuroprotective in Huntington's disease mice. *Mol. Neurodegener.* **2010**, *5*, 18. [CrossRef]
39. Coppen, E.M.; Roos, R.A. Current Pharmacological Approaches to Reduce Chorea in Huntington's Disease. *Drugs* **2017**, *77*, 29–46. [CrossRef]
40. de Tommaso, M.; Serpino, C.; Sciruicchio, V. Management of Huntington's disease: Role of tetrabenazine. *Ther. Clin. Risk Manag.* **2011**, *7*, 123–129. [CrossRef]
41. Claassen, D.O.; Carroll, B.; De Boer, L.M.; Wu, E.; Ayyagari, R.; Gandhi, S.; Stamler, D. Indirect tolerability comparison of Deutetrabenazine and Tetrabenazine for Huntington disease. *J. Clin. Mov. Disord.* **2017**, *4*, 3. [CrossRef]
42. Chen, M.; Ona, V.O.; Li, M.; Ferrante, R.J.; Fink, K.B.; Zhu, S.; Bian, J.; Guo, L.; Farrell, L.A.; Hersch, S.M.; et al. Minocycline inhibits caspase-1 and caspase-3 expression and delays mortality in a transgenic mouse model of Huntington disease. *Nat. Med.* **2000**, *6*, 797–801. [CrossRef] [PubMed]
43. Bonelli, R.M.; Heuberger, C.; Reisecker, F. Minocycline for Huntington's disease: An open label study. *Neurology* **2003**, *60*, 883–884. [CrossRef] [PubMed]
44. Bonelli, R.M.; Hodl, A.K.; Hofmann, P.; Kapfhammer, H.P. Neuroprotection in Huntington's disease: A 2-year study on minocycline. *Int. Clin. Psychopharmacol.* **2004**, *19*, 337–342. [CrossRef] [PubMed]
45. Thomas, M.; Ashizawa, T.; Jankovic, J. Minocycline in Huntington's disease: A pilot study. *Mov. Disord.* **2004**, *19*, 692–695. [CrossRef] [PubMed]

46. Sanchez, I.; Mahlke, C.; Yuan, J. Pivotal role of oligomerization in expanded polyglutamine neurodegenerative disorders. *Nature* **2003**, *421*, 373–379. [CrossRef]
47. Frid, P.; Anisimov, S.V.; Popovic, N. Congo red and protein aggregation in neurodegenerative diseases. *Brain Res. Rev.* **2007**, *53*, 135–160. [CrossRef]
48. McGowan, D.P.; van Roon-Mom, W.; Holloway, H.; Bates, G.P.; Mangiarini, L.; Cooper, G.J.; Faull, R.L.; Snell, R.G. Amyloid-like inclusions in Huntington's disease. *Neuroscience* **2000**, *100*, 677–680. [CrossRef]
49. Tanaka, M.; Machida, Y.; Niu, S.; Ikeda, T.; Jana, N.R.; Doi, H.; Kurosawa, M.; Nekooki, M.; Nukina, N. Trehalose alleviates polyglutamine-mediated pathology in a mouse model of Huntington disease. *Nat. Med.* **2004**, *10*, 148–154. [CrossRef]
50. Lee, H.J.; Yoon, Y.S.; Lee, S.J. Mechanism of neuroprotection by trehalose: Controversy surrounding autophagy induction. *Cell Death Dis.* **2018**, *9*, 712. [CrossRef]
51. Fernandez-Estevez, M.A.; Casarejos, M.J.; Lopez Sendon, J.; Garcia Caldentey, J.; Ruiz, C.; Gomez, A.; Perucho, J.; de Yebenes, J.G.; Mena, M.A. Trehalose reverses cell malfunction in fibroblasts from normal and Huntington's disease patients caused by proteosome inhibition. *PLoS ONE* **2014**, *9*, e90202. [CrossRef]
52. Wang, N.; Lu, X.H.; Sandoval, S.V.; Yang, X.W. An independent study of the preclinical efficacy of C2-8 in the R6/2 transgenic mouse model of Huntington's disease. *J. Huntington's Dis.* **2013**, *2*, 443–451. [CrossRef] [PubMed]
53. Chopra, V.; Fox, J.H.; Lieberman, G.; Dorsey, K.; Matson, W.; Waldmeier, P.; Housman, D.E.; Kazantsev, A.; Young, A.B.; Hersch, S.; et al. A small-molecule therapeutic lead for Huntington's disease: Preclinical pharmacology and efficacy of C2-8 in the R6/2 transgenic mouse. *Proc. Natl. Acad. Sci. USA* **2007**, *104*, 16685–16689. [CrossRef] [PubMed]
54. Ravikumar, B.; Vacher, C.; Berger, Z.; Davies, J.E.; Luo, S.; Oroz, L.G.; Scaravilli, F.; Easton, D.F.; Duden, R.; O'Kane, C.J.; et al. Inhibition of mTOR induces autophagy and reduces toxicity of polyglutamine expansions in fly and mouse models of Huntington disease. *Nat. Genet.* **2004**, *36*, 585–595. [CrossRef] [PubMed]
55. Saxton, R.A.; Sabatini, D.M. mTOR Signaling in Growth, Metabolism, and Disease. *Cell* **2017**, *168*, 960–976. [CrossRef]
56. Pryor, W.M.; Biagioli, M.; Shahani, N.; Swarnkar, S.; Huang, W.C.; Page, D.T.; MacDonald, M.E.; Subramaniam, S. Huntingtin promotes mTORC1 signaling in the pathogenesis of Huntington's disease. *Sci. Signal.* **2014**, *7*, ra103. [CrossRef]
57. Hersch, S.M.; Schifitto, G.; Oakes, D.; Bredlau, A.L.; Meyers, C.M.; Nahin, R.; Rosas, H.D.; Huntington Study Group CREST-E Investigators and Coordinators. The CREST-E study of creatine for Huntington disease: A randomized controlled trial. *Neurology* **2017**, *89*, 594–601. [CrossRef]
58. McGarry, A.; McDermott, M.; Kieburtz, K.; de Blieck, E.A.; Beal, F.; Marder, K.; Ross, C.; Shoulson, I.; Gilbert, P.; Mallonee, W.M.; et al. A randomized, double-blind, placebo-controlled trial of coenzyme Q10 in Huntington disease. *Neurology* **2017**, *88*, 152–159. [CrossRef]
59. Ferreira, J.J.; Rosser, A.; Craufurd, D.; Squitieri, F.; Mallard, N.; Landwehrmeyer, B. Ethyl-eicosapentaenoic acid treatment in Huntington's disease: A placebo-controlled clinical trial. *Mov. Disord.* **2015**, *30*, 1426–1429. [CrossRef]
60. Mao, Z.; Choo, Y.S.; Lesort, M. Cystamine and cysteamine prevent 3-NP-induced mitochondrial depolarization of Huntington's disease knock-in striatal cells. *Eur. J. Neurosci.* **2006**, *23*, 1701–1710. [CrossRef]
61. Gohil, V.M.; Offner, N.; Walker, J.A.; Sheth, S.A.; Fossale, E.; Gusella, J.F.; MacDonald, M.E.; Neri, C.; Mootha, V.K. Meclizine is neuroprotective in models of Huntington's disease. *Hum. Mol. Genet.* **2011**, *20*, 294–300. [CrossRef]
62. Gardian, G.; Browne, S.E.; Choi, D.K.; Klivenyi, P.; Gregorio, J.; Kubilus, J.K.; Ryu, H.; Langley, B.; Ratan, R.R.; Ferrante, R.J.; et al. Neuroprotective effects of phenylbutyrate in the N171-82Q transgenic mouse model of Huntington's disease. *J. Biol. Chem.* **2005**, *280*, 556–563. [CrossRef] [PubMed]
63. Thomas, E.A.; Coppola, G.; Desplats, P.A.; Tang, B.; Soragni, E.; Burnett, R.; Gao, F.; Fitzgerald, K.M.; Borok, J.F.; Herman, D.; et al. The HDAC inhibitor 4b ameliorates the disease phenotype and transcriptional abnormalities in Huntington's disease transgenic mice. *Proc. Natl. Acad. Sci. USA* **2008**, *105*, 15564–15569. [CrossRef] [PubMed]

64. Hockly, E.; Richon, V.M.; Woodman, B.; Smith, D.L.; Zhou, X.; Rosa, E.; Sathasivam, K.; Ghazi-Noori, S.; Mahal, A.; Lowden, P.A.; et al. Suberoylanilide hydroxamic acid, a histone deacetylase inhibitor, ameliorates motor deficits in a mouse model of Huntington's disease. *Proc. Natl. Acad. Sci. USA* **2003**, *100*, 2041–2046. [CrossRef] [PubMed]

65. Ryu, H.; Lee, J.; Hagerty, S.W.; Soh, B.Y.; McAlpin, S.E.; Cormier, K.A.; Smith, K.M.; Ferrante, R.J. ESET/SETDB1 gene expression and histone H3 (K9) trimethylation in Huntington's disease. *Proc. Natl. Acad. Sci. USA* **2006**, *103*, 19176–19181. [CrossRef]

66. Southwell, A.L.; Ko, J.; Patterson, P.H. Intrabody gene therapy ameliorates motor, cognitive, and neuropathological symptoms in multiple mouse models of Huntington's disease. *J. Neurosci.* **2009**, *29*, 13589–13602. [CrossRef]

67. Garriga-Canut, M.; Agustin-Pavon, C.; Herrmann, F.; Sanchez, A.; Dierssen, M.; Fillat, C.; Isalan, M. Synthetic zinc finger repressors reduce mutant huntingtin expression in the brain of R6/2 mice. *Proc. Natl. Acad. Sci. USA* **2012**, *109*, E3136–E3145. [CrossRef]

68. Cox, D.B.; Platt, R.J.; Zhang, F. Therapeutic genome editing: Prospects and challenges. *Nat. Med.* **2015**, *21*, 121–131. [CrossRef]

69. Yang, S.; Chang, R.; Yang, H.; Zhao, T.; Hong, Y.; Kong, H.E.; Sun, X.; Qin, Z.; Jin, P.; Li, S.; et al. CRISPR/Cas9-mediated gene editing ameliorates neurotoxicity in mouse model of Huntington's disease. *J. Clin. Investg.* **2017**, *127*, 2719–2724. [CrossRef]

70. Kordasiewicz, H.B.; Stanek, L.M.; Wancewicz, E.V.; Mazur, C.; McAlonis, M.M.; Pytel, K.A.; Artates, J.W.; Weiss, A.; Cheng, S.H.; Shihabuddin, L.S.; et al. Sustained therapeutic reversal of Huntington's disease by transient repression of huntingtin synthesis. *Neuron* **2012**, *74*, 1031–1044. [CrossRef]

71. Bennett, C.F.; Swayze, E.E. RNA targeting therapeutics: Molecular mechanisms of antisense oligonucleotides as a therapeutic platform. *Annu. Rev. Pharmacol. Toxicol.* **2010**, *50*, 259–293. [CrossRef]

72. Franich, N.R.; Fitzsimons, H.L.; Fong, D.M.; Klugmann, M.; During, M.J.; Young, D. AAV vector-mediated RNAi of mutant huntingtin expression is neuroprotective in a novel genetic rat model of Huntington's disease. *Mol. Ther.* **2008**, *16*, 947–956. [CrossRef] [PubMed]

73. Stanek, L.M.; Sardi, S.P.; Mastis, B.; Richards, A.R.; Treleaven, C.M.; Taksir, T.; Misra, K.; Cheng, S.H.; Shihabuddin, L.S. Silencing mutant huntingtin by adeno-associated virus-mediated RNA interference ameliorates disease manifestations in the YAC128 mouse model of Huntington's disease. *Hum. Gene Ther.* **2014**, *25*, 461–474. [CrossRef] [PubMed]

74. Samaranch, L.; Blits, B.; San Sebastian, W.; Hadaczek, P.; Bringas, J.; Sudhakar, V.; Macayan, M.; Pivirotto, P.J.; Petry, H.; Bankiewicz, K.S.; et al. MR-guided parenchymal delivery of adeno-associated viral vector serotype 5 in non-human primate brain. *Gene Ther.* **2017**, *24*, 253–261. [CrossRef] [PubMed]

75. Wang, H.; Lim, P.J.; Yin, C.; Rieckher, M.; Vogel, B.E.; Monteiro, M.J. Suppression of polyglutamine-induced toxicity in cell and animal models of Huntington's disease by ubiquilin. *Hum. Mol. Genet.* **2006**, *15*, 1025–1041. [CrossRef]

76. Safren, N.; El Ayadi, A.; Chang, L.; Terrillion, C.E.; Gould, T.D.; Boehning, D.F.; Monteiro, M.J. Ubiquilin-1 overexpression increases the lifespan and delays accumulation of Huntingtin aggregates in the R6/2 mouse model of Huntington's disease. *PLoS ONE* **2014**, *9*, e87513. [CrossRef]

77. Thulasiraman, V.; Yang, C.F.; Frydman, J. In vivo newly translated polypeptides are sequestered in a protected folding environment. *EMBO J.* **1999**, *18*, 85–95. [CrossRef]

78. Kalisman, N.; Adams, C.M.; Levitt, M. Subunit order of eukaryotic TRiC/CCT chaperonin by cross-linking, mass spectrometry, and combinatorial homology modeling. *Proc. Natl. Acad. Sci. USA* **2012**, *109*, 2884–2889. [CrossRef]

79. Reilmann, R.; Rouzade-Dominguez, M.L.; Saft, C.; Sussmuth, S.D.; Priller, J.; Rosser, A.; Rickards, H.; Schols, L.; Pezous, N.; Gasparini, F.; et al. A randomized, placebo-controlled trial of AFQ056 for the treatment of chorea in Huntington's disease. *Mov. Disord.* **2015**, *30*, 427–431. [CrossRef]

80. Chabrier, P.E.; Auguet, M. Pharmacological properties of BN82451: A novel multitargeting neuroprotective agent. *CNS Drug Rev.* **2007**, *13*, 317–332. [CrossRef]

81. Klivenyi, P.; Ferrante, R.J.; Gardian, G.; Browne, S.; Chabrier, P.E.; Beal, M.F. Increased survival and neuroprotective effects of BN82451 in a transgenic mouse model of Huntington's disease. *J. Neurochem.* **2003**, *86*, 267–272. [CrossRef]

82. Graham, R.K.; Deng, Y.; Slow, E.J.; Haigh, B.; Bissada, N.; Lu, G.; Pearson, J.; Shehadeh, J.; Bertram, L.; Murphy, Z.; et al. Cleavage at the caspase-6 site is required for neuronal dysfunction and degeneration due to mutant huntingtin. *Cell* **2006**, *125*, 1179–1191. [CrossRef] [PubMed]

83. Wellington, C.L.; Ellerby, L.M.; Hackam, A.S.; Margolis, R.L.; Trifiro, M.A.; Singaraja, R.; McCutcheon, K.; Salvesen, G.S.; Propp, S.S.; Bromm, M.; et al. Caspase cleavage of gene products associated with triplet expansion disorders generates truncated fragments containing the polyglutamine tract. *J. Biol. Chem.* **1998**, *273*, 9158–9167. [CrossRef] [PubMed]

84. Marder, K.; Gu, Y.; Eberly, S.; Tanner, C.M.; Scarmeas, N.; Oakes, D.; Shoulson, I.; Huntington Study Group, P.I. Relationship of Mediterranean diet and caloric intake to phenoconversion in Huntington disease. *JAMA Neurol.* **2013**, *70*, 1382–1388. [CrossRef] [PubMed]

85. Ryu, H.; Rosas, H.D.; Hersch, S.M.; Ferrante, R.J. The therapeutic role of creatine in Huntington's disease. *Pharmacol. Ther.* **2005**, *108*, 193–207. [CrossRef] [PubMed]

86. Hersch, S.M.; Gevorkian, S.; Marder, K.; Moskowitz, C.; Feigin, A.; Cox, M.; Como, P.; Zimmerman, C.; Lin, M.; Zhang, L.; et al. Creatine in Huntington disease is safe, tolerable, bioavailable in brain and reduces serum 8OH2'dG. *Neurology* **2006**, *66*, 250–252. [CrossRef]

87. Rosas, H.D.; Doros, G.; Gevorkian, S.; Malarick, K.; Reuter, M.; Coutu, J.P.; Triggs, T.D.; Wilkens, P.J.; Matson, W.; Salat, D.H.; et al. PRECREST: A phase II prevention and biomarker trial of creatine in at-risk Huntington disease. *Neurology* **2014**, *82*, 850–857. [CrossRef]

88. Verbessem, P.; Lemiere, J.; Eijnde, B.O.; Swinnen, S.; Vanhees, L.; Van Leemputte, M.; Hespel, P.; Dom, R. Creatine supplementation in Huntington's disease: A placebo-controlled pilot trial. *Neurology* **2003**, *61*, 925–930. [CrossRef]

89. Andrich, J.; Saft, C.; Gerlach, M.; Schneider, B.; Arz, A.; Kuhn, W.; Muller, T. Coenzyme Q10 serum levels in Huntington's disease. *J. Neural Transm. Suppl.* **2004**, *68*, 111–116. [CrossRef]

90. Ferrante, R.J.; Andreassen, O.A.; Dedeoglu, A.; Ferrante, K.L.; Jenkins, B.G.; Hersch, S.M.; Beal, M.F. Therapeutic effects of coenzyme Q10 and remacemide in transgenic mouse models of Huntington's disease. *J. Neurosci.* **2002**, *22*, 1592–1599. [CrossRef]

91. Yang, L.; Calingasan, N.Y.; Wille, E.J.; Cormier, K.; Smith, K.; Ferrante, R.J.; Beal, M.F. Combination therapy with coenzyme Q10 and creatine produces additive neuroprotective effects in models of Parkinson's and Huntington's diseases. *J. Neurochem.* **2009**, *109*, 1427–1439. [CrossRef]

92. Jump, D.B. The biochemistry of n-3 polyunsaturated fatty acids. *J. Biol. Chem.* **2002**, *277*, 8755–8758. [CrossRef] [PubMed]

93. Lonergan, P.E.; Martin, D.S.; Horrobin, D.F.; Lynch, M.A. Neuroprotective effect of eicosapentaenoic acid in hippocampus of rats exposed to gamma-irradiation. *J. Biol. Chem.* **2002**, *277*, 20804–20811. [CrossRef] [PubMed]

94. Martin, D.S.; Lonergan, P.E.; Boland, B.; Fogarty, M.P.; Brady, M.; Horrobin, D.F.; Campbell, V.A.; Lynch, M.A. Apoptotic changes in the aged brain are triggered by interleukin-1beta-induced activation of p38 and reversed by treatment with eicosapentaenoic acid. *J. Biol. Chem.* **2002**, *277*, 34239–34246. [CrossRef]

95. Puri, B.K.; Leavitt, B.R.; Hayden, M.R.; Ross, C.A.; Rosenblatt, A.; Greenamyre, J.T.; Hersch, S.; Vaddadi, K.S.; Sword, A.; Horrobin, D.F.; et al. Ethyl-EPA in Huntington disease: A double-blind, randomized, placebo-controlled trial. *Neurology* **2005**, *65*, 286–292. [CrossRef] [PubMed]

96. Huntington Study Group TREND-HD Investigators. Randomized controlled trial of ethyl-eicosapentaenoic acid in Huntington disease: The TREND-HD study. *Arch. Neurol.* **2008**, *65*, 1582–1589. [CrossRef]

97. Hogarth, P.; Lovrecic, L.; Krainc, D. Sodium phenylbutyrate in Huntington's disease: A dose-finding study. *Mov. Disord.* **2007**, *22*, 1962–1964. [CrossRef] [PubMed]

98. Stack, E.C.; Del Signore, S.J.; Luthi-Carter, R.; Soh, B.Y.; Goldstein, D.R.; Matson, S.; Goodrich, S.; Markey, A.L.; Cormier, K.; Hagerty, S.W.; et al. Modulation of nucleosome dynamics in Huntington's disease. *Hum. Mol. Genet.* **2007**, *16*, 1164–1175. [CrossRef]

99. Lee, J.; Hwang, Y.J.; Kim, K.Y.; Kowall, N.W.; Ryu, H. Epigenetic mechanisms of neurodegeneration in Huntington's disease. *Neurotherapeutics* **2013**, *10*, 664–676. [CrossRef]

100. Sah, D.W.; Aronin, N. Oligonucleotide therapeutic approaches for Huntington disease. *J. Clin. Investg.* **2011**, *121*, 500–507. [CrossRef]

101. Miniarikova, J.; Evers, M.M.; Konstantinova, P. Translation of MicroRNA-Based Huntingtin-Lowering Therapies from Preclinical Studies to the Clinic. *Mol. Ther.* **2018**, *26*, 947–962. [CrossRef]

102. Wild, E.J.; Tabrizi, S.J. Therapies targeting DNA and RNA in Huntington's disease. *Lancet Neurol.* **2017**, *16*, 837–847. [CrossRef]

103. Hu, J.; Matsui, M.; Corey, D.R. Allele-selective inhibition of mutant huntingtin by peptide nucleic acid-peptide conjugates, locked nucleic acid, and small interfering RNA. *Ann. N. Y. Acad. Sci.* **2009**, *1175*, 24–31. [CrossRef] [PubMed]

104. Boudreau, R.L.; McBride, J.L.; Martins, I.; Shen, S.; Xing, Y.; Carter, B.J.; Davidson, B.L. Nonallele-specific silencing of mutant and wild-type huntingtin demonstrates therapeutic efficacy in Huntington's disease mice. *Mol. Ther.* **2009**, *17*, 1053–1063. [CrossRef] [PubMed]

105. Drouet, V.; Perrin, V.; Hassig, R.; Dufour, N.; Auregan, G.; Alves, S.; Bonvento, G.; Brouillet, E.; Luthi-Carter, R.; Hantraye, P.; et al. Sustained effects of nonallele-specific Huntingtin silencing. *Ann. Neurol.* **2009**, *65*, 276–285. [CrossRef] [PubMed]

106. Marelli, C.; Maschat, F. The P42 peptide and Peptide-based therapies for Huntington's disease. *Orphanet J. Rare Dis.* **2016**, *11*, 24. [CrossRef] [PubMed]

107. Malankhanova, T.B.; Malakhova, A.A.; Medvedev, S.P.; Zakian, S.M. Modern Genome Editing Technologies in Huntington's Disease Research. *J. Huntingt. Dis.* **2017**, *6*, 19–31. [CrossRef] [PubMed]

108. Vachey, G.; Deglon, N. CRISPR/Cas9-Mediated Genome Editing for Huntington's Disease. *Methods Mol. Biol.* **2018**, *1780*, 463–481. [CrossRef]

109. Monteys, A.M.; Ebanks, S.A.; Keiser, M.S.; Davidson, B.L. CRISPR/Cas9 Editing of the Mutant Huntingtin Allele In Vitro and In Vivo. *Mol. Ther.* **2017**, *25*, 12–23. [CrossRef]

110. Dabrowska, M.; Olejniczak, M. Gene Therapy for Huntington's Disease Using Targeted Endonucleases. *Methods Mol. Biol.* **2020**, *2056*, 269–284. [CrossRef]

111. Ekman, F.K.; Ojala, D.S.; Adil, M.M.; Lopez, P.A.; Schaffer, D.V.; Gaj, T. CRISPR-Cas9-Mediated Genome Editing Increases Lifespan and Improves Motor Deficits in a Huntington's Disease Mouse Model. *Mol. Ther. Nucleic Acids* **2019**, *17*, 829–839. [CrossRef]

112. Larrouy, B.; Blonski, C.; Boiziau, C.; Stuer, M.; Moreau, S.; Shire, D.; Toulme, J.J. RNase H-mediated inhibition of translation by antisense oligodeoxyribonucleotides: Use of backbone modification to improve specificity. *Gene* **1992**, *121*, 189–194. [CrossRef]

113. Crooke, S.T.; Baker, B.F.; Kwoh, T.J.; Cheng, W.; Schulz, D.J.; Xia, S.; Salgado, N.; Bui, H.H.; Hart, C.E.; Burel, S.A.; et al. Integrated Safety Assessment of 2'-O-Methoxyethyl Chimeric Antisense Oligonucleotides in NonHuman Primates and Healthy Human Volunteers. *Mol. Ther.* **2016**, *24*, 1771–1782. [CrossRef] [PubMed]

114. Stanek, L.M.; Yang, W.; Angus, S.; Sardi, P.S.; Hayden, M.R.; Hung, G.H.; Bennett, C.F.; Cheng, S.H.; Shihabuddin, L.S. Antisense oligonucleotide-mediated correction of transcriptional dysregulation is correlated with behavioral benefits in the YAC128 mouse model of Huntington's disease. *J. Huntingt. Dis.* **2013**, *2*, 217–228. [CrossRef] [PubMed]

115. Carroll, J.B.; Warby, S.C.; Southwell, A.L.; Doty, C.N.; Greenlee, S.; Skotte, N.; Hung, G.; Bennett, C.F.; Freier, S.M.; Hayden, M.R.; et al. Potent and selective antisense oligonucleotides targeting single-nucleotide polymorphisms in the Huntington disease gene / allele-specific silencing of mutant huntingtin. *Mol. Ther.* **2011**, *19*, 2178–2185. [CrossRef]

116. Hammond, S.M.; Hazell, G.; Shabanpoor, F.; Saleh, A.F.; Bowerman, M.; Sleigh, J.N.; Meijboom, K.E.; Zhou, H.; Muntoni, F.; Talbot, K.; et al. Systemic peptide-mediated oligonucleotide therapy improves long-term survival in spinal muscular atrophy. *Proc. Natl. Acad. Sci. USA* **2016**, *113*, 10962–10967. [CrossRef]

117. Tabrizi, S.J.; Leavitt, B.R.; Landwehrmeyer, G.B.; Wild, E.J.; Saft, C.; Barker, R.A.; Blair, N.F.; Craufurd, D.; Priller, J.; Rickards, H.; et al. Targeting Huntingtin Expression in Patients with Huntington's Disease. *N. Engl. J. Med.* **2019**, *380*, 2307–2316. [CrossRef]

118. Cho, I.K.; Hunter, C.E.; Ye, S.; Pongos, A.L.; Chan, A.W.S. Combination of stem cell and gene therapy ameliorates symptoms in Huntington's disease mice. *NPJ Regen. Med.* **2019**, *4*, 7. [CrossRef]

119. Aguiar, S.; van der Gaag, B.; Cortese, F.A.B. RNAi mechanisms in Huntington's disease therapy: siRNA versus shRNA. *Transl. Neurodegener.* **2017**, *6*, 30. [CrossRef]

120. Naryshkin, N.A.; Weetall, M.; Dakka, A.; Narasimhan, J.; Zhao, X.; Feng, Z.; Ling, K.K.; Karp, G.M.; Qi, H.; Woll, M.G.; et al. Motor neuron disease. SMN2 splicing modifiers improve motor function and longevity in mice with spinal muscular atrophy. *Science* **2014**, *345*, 688–693. [CrossRef]

121. Ratni, H.; Karp, G.M.; Weetall, M.; Naryshkin, N.A.; Paushkin, S.V.; Chen, K.S.; McCarthy, K.D.; Qi, H.; Turpoff, A.; Woll, M.G.; et al. Specific Correction of Alternative Survival Motor Neuron 2 Splicing by Small Molecules: Discovery of a Potential Novel Medicine to Treat Spinal Muscular Atrophy. *J. Med. Chem.* **2016**, *59*, 6086–6100. [CrossRef]

122. Kletzl, H.; Marquet, A.; Gunther, A.; Tang, W.; Heuberger, J.; Groeneveld, G.J.; Birkhoff, W.; Mercuri, E.; Lochmuller, H.; Wood, C.; et al. The oral splicing modifier RG7800 increases full length survival of motor neuron 2 mRNA and survival of motor neuron protein: Results from trials in healthy adults and patients with spinal muscular atrophy. *Neuromuscul. Disord.* **2019**, *29*, 21–29. [CrossRef] [PubMed]

123. Caron, N.S.; Dorsey, E.R.; Hayden, M.R. Therapeutic approaches to Huntington disease: From the bench to the clinic. *Nat. Rev. Drug Discov.* **2018**, *17*, 729–750. [CrossRef]

124. Pattison, L.R.; Kotter, M.R.; Fraga, D.; Bonelli, R.M. Apoptotic cascades as possible targets for inhibiting cell death in Huntington's disease. *J. Neurol.* **2006**, *253*, 1137–1142. [CrossRef] [PubMed]

125. Sofi, F.; Macchi, C.; Casini, A. Mediterranean Diet and Minimizing Neurodegeneration. *Curr. Nutr. Rep.* **2013**, *2*, 75–80. [CrossRef]

126. Morris, M.C.; Tangney, C.C.; Wang, Y.; Sacks, F.M.; Bennett, D.A.; Aggarwal, N.T. MIND diet associated with reduced incidence of Alzheimer's disease. *Alzheimers Dement.* **2015**, *11*, 1007–1014. [CrossRef]

127. Ehrnhoefer, D.E.; Martin, D.D.O.; Schmidt, M.E.; Qiu, X.; Ladha, S.; Caron, N.S.; Skotte, N.H.; Nguyen, Y.T.N.; Vaid, K.; Southwell, A.L.; et al. Preventing mutant huntingtin proteolysis and intermittent fasting promote autophagy in models of Huntington disease. *Acta Neuropathol. Commun.* **2018**, *6*, 16. [CrossRef]

128. Borrell-Pages, M.; Canals, J.M.; Cordelieres, F.P.; Parker, J.A.; Pineda, J.R.; Grange, G.; Bryson, E.A.; Guillermier, M.; Hirsch, E.; Hantraye, P.; et al. Cystamine and cysteamine increase brain levels of BDNF in Huntington disease via HSJ1b and transglutaminase. *J. Clin. Investg.* **2006**, *116*, 1410–1424. [CrossRef]

129. Menalled, L.B.; Kudwa, A.E.; Oakeshott, S.; Farrar, A.; Paterson, N.; Filippov, I.; Miller, S.; Kwan, M.; Olsen, M.; Beltran, J.; et al. Genetic deletion of transglutaminase 2 does not rescue the phenotypic deficits observed in R6/2 and zQ175 mouse models of Huntington's disease. *PLoS ONE* **2014**, *9*, e99520. [CrossRef]

130. Waters, S.; Tedroff, J.; Ponten, H.; Klamer, D.; Sonesson, C.; Waters, N. Pridopidine: Overview of Pharmacology and Rationale for its Use in Huntington's Disease. *J. Hunt. Dis.* **2018**, *7*, 1–16. [CrossRef]

131. Ryskamp, D.; Wu, L.; Wu, J.; Kim, D.; Rammes, G.; Geva, M.; Hayden, M.; Bezprozvanny, I. Pridopidine stabilizes mushroom spines in mouse models of Alzheimer's disease by acting on the sigma-1 receptor. *Neurobiol. Dis.* **2019**, *124*, 489–504. [CrossRef]

132. de Yebenes, J.G.; Landwehrmeyer, B.; Squitieri, F.; Reilmann, R.; Rosser, A.; Barker, R.A.; Saft, C.; Magnet, M.K.; Sword, A.; Rembratt, A.; et al. Pridopidine for the treatment of motor function in patients with Huntington's disease (MermaiHD): A phase 3, randomised, double-blind, placebo-controlled trial. *Lancet Neurol.* **2011**, *10*, 1049–1057. [CrossRef]

133. Reilmann, R.; McGarry, A.; Grachev, I.D.; Savola, J.M.; Borowsky, B.; Eyal, E.; Gross, N.; Langbehn, D.; Schubert, R.; Wickenberg, A.T.; et al. Safety and efficacy of pridopidine in patients with Huntington's disease (PRIDE-HD): A phase 2, randomised, placebo-controlled, multicentre, dose-ranging study. *Lancet Neurol.* **2019**, *18*, 165–176. [CrossRef]

134. Geva, M.; Kusko, R.; Soares, H.; Fowler, K.D.; Birnberg, T.; Barash, S.; Wagner, A.M.; Fine, T.; Lysaght, A.; Weiner, B.; et al. Pridopidine activates neuroprotective pathways impaired in Huntington Disease. *Hum. Mol. Genet.* **2016**, *25*, 3975–3987. [CrossRef] [PubMed]

135. Pineda, J.R.; Canals, J.M.; Bosch, M.; Adell, A.; Mengod, G.; Artigas, F.; Ernfors, P.; Alberch, J. Brain-derived neurotrophic factor modulates dopaminergic deficits in a transgenic mouse model of Huntington's disease. *J. Neurochem.* **2005**, *93*, 1057–1068. [CrossRef] [PubMed]

136. Squitieri, F.; Di Pardo, A.; Favellato, M.; Amico, E.; Maglione, V.; Frati, L. Pridopidine, a dopamine stabilizer, improves motor performance and shows neuroprotective effects in Huntington disease R6/2 mouse model. *J. Cell. Mol. Med.* **2015**, *19*, 2540–2548. [CrossRef]

137. Huntington Study Group HART Investigators. A randomized, double-blind, placebo-controlled trial of pridopidine in Huntington's disease. *Mov. Disord.* **2013**, *28*, 1407–1415. [CrossRef] [PubMed]

138. Adanyeguh, I.M.; Rinaldi, D.; Henry, P.G.; Caillet, S.; Valabregue, R.; Durr, A.; Mochel, F. Triheptanoin improves brain energy metabolism in patients with Huntington disease. *Neurology* **2015**, *84*, 490–495. [CrossRef]

139. Kieburtz, K.; McDermott, M.P.; Voss, T.S.; Corey-Bloom, J.; Deuel, L.M.; Dorsey, E.R.; Factor, S.; Geschwind, M.D.; Hodgeman, K.; Kayson, E.; et al. A randomized, placebo-controlled trial of latrepirdine in Huntington disease. *Arch. Neurol.* **2010**, *67*, 154–160. [CrossRef]

140. Blanpied, T.A.; Clarke, R.J.; Johnson, J.W. Amantadine inhibits NMDA receptors by accelerating channel closure during channel block. *J. Neurosci* **2005**, *25*, 3312–3322. [CrossRef]

141. Hosenbocus, S.; Chahal, R. Amantadine: A review of use in child and adolescent psychiatry. *J. Can. Acad Child. Adolesc. Psychiatry* **2013**, *22*, 55–60.

142. Verhagen Metman, L.; Morris, M.J.; Farmer, C.; Gillespie, M.; Mosby, K.; Wuu, J.; Chase, T.N. Huntington's disease: A randomized, controlled trial using the NMDA-antagonist amantadine. *Neurology* **2002**, *59*, 694–699. [CrossRef] [PubMed]

143. Goldsmith, D.R.; Wagstaff, A.J.; Ibbotson, T.; Perry, C.M. Lamotrigine: A review of its use in bipolar disorder. *Drugs* **2003**, *63*, 2029–2050. [CrossRef] [PubMed]

144. Huang, Y.Y.; Liu, Y.C.; Lee, C.T.; Lin, Y.C.; Wang, M.L.; Yang, Y.P.; Chang, K.Y.; Chiou, S.H. Revisiting the Lamotrigine-Mediated Effect on Hippocampal GABAergic Transmission. *Int. J. Mol. Sci.* **2016**, *17*, 1191. [CrossRef] [PubMed]

145. Shen, Y.C. Lamotrigine in motor and mood symptoms of Huntington's disease. *World J. Biol. Psychiatry* **2008**, *9*, 147–149. [CrossRef] [PubMed]

146. Sussmuth, S.D.; Haider, S.; Landwehrmeyer, G.B.; Farmer, R.; Frost, C.; Tripepi, G.; Andersen, C.A.; Di Bacco, M.; Lamanna, C.; Diodato, E.; et al. An exploratory double-blind, randomized clinical trial with selisistat, a SirT1 inhibitor, in patients with Huntington's disease. *Br. J. Clin. Pharmacol.* **2015**, *79*, 465–476. [CrossRef] [PubMed]

147. Reilmann, R.; Squitieri, F.; Priller, J.; Saft, C.; Mariotti, C.; Suessmuth, S.; Nemeth, A.; Tabrizi, S.; Quarrell, O.; Craufurd, D.; et al. Safety and Tolerability of Selisistat for the Treatment of Huntington's Disease: Results from a Randomized, Double-Blind, Placebo-Controlled Phase II Trial (S47.004). *Neurology* **2014**, *82*, S47.004. [CrossRef]

148. Keene, C.D.; Rodrigues, C.M.; Eich, T.; Chhabra, M.S.; Steer, C.J.; Low, W.C. Tauroursodeoxycholic acid, a bile acid, is neuroprotective in a transgenic animal model of Huntington's disease. *Proc. Natl. Acad. Sci. USA* **2002**, *99*, 10671–10676. [CrossRef]

149. Ehrnhoefer, D.E.; Caron, N.S.; Deng, Y.; Qiu, X.; Tsang, M.; Hayden, M.R. Laquinimod decreases Bax expression and reduces caspase-6 activation in neurons. *Exp. Neurol.* **2016**, *283*, 121–128. [CrossRef]

150. Garcia-Miralles, M.; Yusof, N.; Tan, J.Y.; Radulescu, C.I.; Sidik, H.; Tan, L.J.; Belinson, H.; Zach, N.; Hayden, M.R.; Pouladi, M.A.; et al. Laquinimod Treatment Improves Myelination Deficits at the Transcriptional and Ultrastructural Levels in the YAC128 Mouse Model of Huntington Disease. *Mol. Neurobiol.* **2019**, *56*, 4464–4478. [CrossRef]

151. Garcia-Miralles, M.; Hong, X.; Tan, L.J.; Caron, N.S.; Huang, Y.; To, X.V.; Lin, R.Y.; Franciosi, S.; Papapetropoulos, S.; Hayardeny, L.; et al. Laquinimod rescues striatal, cortical and white matter pathology and results in modest behavioural improvements in the YAC128 model of Huntington disease. *Sci. Rep.* **2016**, *6*, 31652. [CrossRef]

152. Mazarei, G.; Leavitt, B.R. Indoleamine 2,3 Dioxygenase as a Potential Therapeutic Target in Huntington's Disease. *J. Huntingt. Dis.* **2015**, *4*, 109–118. [CrossRef] [PubMed]

153. Giorgini, F.; Moller, T.; Kwan, W.; Zwilling, D.; Wacker, J.L.; Hong, S.; Tsai, L.C.; Cheah, C.S.; Schwarcz, R.; Guidetti, P.; et al. Histone deacetylase inhibition modulates kynurenine pathway activation in yeast, microglia, and mice expressing a mutant huntingtin fragment. *J. Biol. Chem.* **2008**, *283*, 7390–7400. [CrossRef] [PubMed]

MDPI

St. Alban-Anlage 66

4052 Basel

Switzerland

Tel. +41 61 683 77 34

Fax +41 61 302 89 18

www.mdpi.com

Brain Sciences Editorial Office

E-mail: brainsci@mdpi.com

www.mdpi.com/journal/brainsci

* 9 7 8 3 0 3 9 4 3 8 1 1 2 *